A YEAR
in the
MAINE WOODS

ADDISON-WESLEY PUBLISHING COMPANY

Reading, Massachusetts   Menlo Park, California   New York
Don Mills, Ontario   Wokingham, England   Amsterdam   Bonn
Sydney   Singapore   Tokyo   Madrid   San Juan
Paris   Seoul   Milan   Mexico City   Taipei

A YEAR

in the

MAINE
WOODS

BERND HEINRICH

Library of Congress Cataloging-in-Publication Data

Heinrich, Bernd, 1940–
    A year in the Maine woods / Bernd Heinrich.
      p. cm.
    ISBN 0-201-62252-1
    ISBN 0-201-48939-2 (pbk.)
    1. Country life—Maine. 2. Natural history—Maine. 3. Maine—
Description and travel. 4. Heinrich, Bernd, 1940– Homes and
haunts—Maine. I. Title.
    F26.H45 1994
    974.1—dc20   94-16534
                                             CIP

Cover design by Jean Seal
Text design by Irving Perkins Associates
Set in 11 point Garamond Light by Pagesetters Inc.

4 5 6 7 8 9 10 11 12—MA—0201009998
Fourth printing, March 1998

For the Adamses—Floyd, Leona, Jim, Bill, and Vernon—
who made the Maine difference.

ACKNOWLEDGMENTS

I thank Sandra Dijkstra, my agent, for making this journey possible, and Bill Patrick, my editor, for his eagle eye in steering it through the woods. Naturally, I also thank Ron and Syndi Gerrish for many a good cup of coffee and other beverages, and for their multiple-use outhouse when I needed it most. Edna York Buchanan provided valuable historical information and I am much indebted to her.

AUTHOR'S NOTE

I don't like doing things by proxy.

By profession I'm a naturalist and a scientist, but I'm also a human being. And whatever else I may aspire to, inevitably, I am what I do. For the past twenty-five years I've been teaching at a university, which means what I also do is fill out forms, read memos, and sit in meetings. Sometimes I apply for grants, and sometimes I write papers, but what I *really* want to do . . . is to be out in the woods.

By the time I was ten years old, I had lived for six years as a refugee in a northern German forest. My family survived on very little. But I had a lot—I had a pet crow, and I could collect beetles. Sometimes lately I have begun to wonder if it is still possible to taste the world up close as I did as a child. I long to see nature again as I did then—fresh, clear, timeless, and shrouded in magic. Would it be possible to rediscover the vividness that I now experience only as brief, tantalizing flashbacks?

Edward Abbey said you must "brew your own beer; kick in your Tee Vee; kill your own beef; build your own cabin and piss off the front porch whenever you bloody well feel like it." I already had had a good start. As a teenager in rural Maine, after we came to America, I had learned hunting, fishing, and trapping in the wilderness. My Maine mentors had long ago taught me to make home brew. I owned a rifle, and I'd already built a log cabin. The rest should be easy. I thought I'd give it a shot.

SUMMER

A TRAVELING COMPANION

The route from Burlington, Vermont, to my cabin is about 200 miles long. Most of the way you travel on Route 2, through the cities of Montpelier and St. Johnsbury, Vermont; through Lancaster and Gorham, New Hampshire, at the foot of the northern Presidentials; and then to Rumford, Maine, past the great Boise Cascade paper mill. When I smell the Rumford mill, I feel I'm almost home. It's only a few more miles to the foot of Mount Blue, where big Webb Lake fills the valley below Mount Tumbledown, Blueberry Mountain, and Little Jackson Mountain, but it's been a long ride for a baby raven.

My friend Chuck Reiss and I got him on May 13, after setting up a ladder that just barely reached to the nest in a recess on a granite cliff. The blackflies attacked us mercilessly, but the birds' parents left when we came. Chuck, who builds houses, climbed easily to the top and I watched his progress, for once quite satisfied to experience the thrill vicariously. I had already done the hard part—getting the permits.

The young bird Chuck brought down was feathered out and already close to adult weight. His wing and tail feathers had a beautiful bluish-purple sheen, but he still had a few wispy tufts of white down on the top of his head. He was clumsy, not yet a flyer, but already a good hopper. He did not appear to be afraid of us as he calmly looked us over with steely light blue eyes. He readily seized the meat that we held in front of him. He either swallowed it right away, or spit it out and flung it violently aside if he did not approve of its taste.

We grabbed him after he tried to hop away, and he made loud, angry, rasping calls like a chain dragged over jagged metal. But he soon settled comfortably into a nest in a cage in my living room, and there he tolerated having his head scratched. Gentle petting made him

3

sit down and close his eyes. The rest of the time he preened a lot and stretched often. Using his thick bill, he played with the leaves, sticks, and grass stems at the edge of his nest. I had already watched him for many hours from a nest-level blind I had built in the top of a red maple tree, and he behaved now much as he had then. He had been the largest of the brood, always alert and playing. I decided to call him Jack.

On the road back from Vermont, I am feeling anxious now about having Jack up front with me in the pickup truck. My first concern is his hyperactivity. I recall the "roadkill" barred owl which once very much revived itself on my back seat before proceeding to use my steering wheel as its perch. A raven is much larger, much more active, and eats more. It's not that I lack sufficient entrees for Jack—chopped squirrel, flattened cedar waxwing—but the speed of his digestion disturbs me. Blueberries pass through in 15 or 20 minutes.

Before leaving, I put a couple of cardboard boxes filled with books and such in the bed of the truck, but the cab was reserved for us. To make the ride as comfortable as possible for Jack, I taped a tree limb as a perch between the two front seats. That way he could sit next to my right ear while still keeping an eye on the road too, possibly looking for roadkills. As I conveniently feed his front end, his droppings should cascade onto the newspapers covering the floor. As a precaution, I taped plastic sheeting over both seats, in case he got restless on his perch.

It was evening when we began our five-hour ride. Jack settled next to my right ear after giving the cab a brief whirl or two. However, as much as I wanted him to snooze, he didn't drop off once the whole way. I often caught the glint in his eye as he kept ever alert, staring at my face. To reward him for this good behavior, I ripped apart the cedar waxwing and handed it to him in little pieces.

His throaty little murmurs told of his contentment and enjoyment of the ride. We talked for the whole while through Vermont and New Hampshire, but it was up to me to keep the conversation going. Without my prompting, Jack often lapsed into silence. I'd tell him that he was a nice bird, that I noticed him, listened to him, and liked him. He'd counter with a soft, low "mm." These "mm's" varied slightly in length and mellowness, reflecting his approval and excitement. Very low, soft and long, almost whispering "mm's" meant that he was very laid-back. Higher, shorter, and louder "mm's" meant heightened awareness and agitation. On occasion, as Jack's anxiety increased, there was also an upward inflection of the sound. When he was mellow, there was a slight downward inflection instead.

Once in a while he'd stop to stretch—with both wings held upward while he crouched down, or with one wing and one leg (toes tightly clenched) extended first to one side and then to the other. I think Jane Fonda would have approved of these exercises, which were accompanied at times by a more assertive or emphatic "krrr" or "kn," meaning "I'm feeling *great!*" Then he might give a little cough before idling away the rest of the time preening his feathers or checking out the passing scenery.

I understood perfectly well what Jack was saying. Mostly, it was to the effect of "I hear you, you're OK; I'm listening, I'm here and A-OK." One can speak volumes with a few little sounds, given the proper context and intonation. With Jack, context is everything. A raven reacts only in the moment, so you always know what he's about.

An hour and a half into the ride, I had made my usual stop in St. Johnsbury at Anthony's Diner, a small place with barstools, featuring the "Woodsman's Burger." I only wanted coffee, but if I took Jack in with me he'd walk all over the customers' plates, steal french fries, check out the coins in the cash register, and generally cause mayhem, so I had to leave him behind. As I closed the cab door, he immediately hopped off his perch and pressed against the window, acting desperate, but I didn't look back, determined to drink a leisurely cup. Besides, absence makes the heart grow fonder.

When I returned 20 minutes later, I found that absence also makes the gut run faster. There were foul streaks all over the seats, and these weren't mere white and scentless droppings. The cedar waxwing had taken the fast route, and wasn't as adequately processed as roadkills should be. He had missed me—the body doesn't lie.

Oh well. When you have a baby, such surprises are part of the bargain.

Once we were off again, Jack quickly resettled on his perch next to my ear.

"You OK, Jack? You glad to see me?"

"Mm*mm.*"

"This is *fun*, huh?"

"Mm*mm.*"

He was *very* glad to see me, and we journeyed on.

When we got to Maine our conversation started to lag, and he had to endure Joan Armatrading and Bruce Springsteen on tape as we drove along the Androscoggin River and finally entered the hills by Dixfield.

ADAMS HILL

Our destination is Adams Hill in western Maine. Once the site of a farm, it is now, except for my small clearing, all forest. When I was a boy, I used to come here with my friends, Phil Potter and Floyd Adams (a descendant of one of the Adamses for whom this hill was named), and we'd see brushy fields and woods overtaking the old apple trees.

People had lived here a long time ago, as they had on hills all around New England. Stone walls that once marked the edges of fields and pastures still braid through these forests like remnants of a lost civilization. You come upon rectangular cellar holes defined by neatly fitted fieldstones and split granite blocks. These barn sills enclose thick white birch, ash, and maple trees. On occasion you still see a bit of rusted barbed wire sticking out of a massive 200-year-old sugar maple that must have once formed a pasture boundary.

My cabin is about a half mile up a steep hill through the brush from a steep winding road, at a place once called Hildreth's Mill that now only exists in the memories of old-timers. Even after a long dry spell has drained the hill so that the path no longer doubles as a small brook, you can't drive there except with a four-wheel-drive vehicle. This path isolates me well from casual visitors, but true friends are not deterred. I like it that way and accordingly have not made great efforts at road improvement.

The cabin walls are of spruce and fir logs that I chopped down with my ax and peeled with a bark spud when they were still full of sap and when the bark still separated easily, like a skin. My former wife, Margaret, and I hired oxen to drag the shiny white logs out into the clearing. Using ropes, levers, and inclines, we hoisted up the logs, which I flattened on two sides by chopping off long, smooth flakes, and then notched at the ends with deep, smooth cuts. Gradually we assembled the structure, over the course of a summer.

It is long after dark when Jack and I arrive, but by now I can walk up the path with my eyes closed. I carry Jack in a cardboard box and he seems relaxed as we have a quiet conversation in the dark. During our first night at the cabin, he gets to stay beside my bed in the box so I can talk to him and reassure him.

Early the next morning, I awake to a wild melee of bird song. Far sweeter than any symphony I could possibly imagine, it comes from all around. The clearing surrounding the cabin grows spirea bushes, and there is a ground cover of ever-green club mosses, close relatives of the giants that produced our coal in the carboniferous ages. There are also maple, spruce, pine and balsam fir saplings, and patches of wild raspberries and blueberries. This profusion of plants that I have been neglectful of pruning in the last several years is now a haven for white-throated sparrows, juncos, chestnut-sided and Nashville warblers, and deep-blue indigo buntings. These birds all sing at intervals, as does the phoebe that has built its nest over my window. I hear the blue jays, rose-breasted grosbeaks, and red-eyed vireos calling from the forest, and I vault out of bed, feeling like a new man.

Looking out the window on this first day in early June, I see a sea of shining green. I had not realized how varied and vibrant green, the color of life, could be. The hue is darkest where you look through several layers of the recently unfurled maple leaves, and it shines with a yellowish tinge where the sun filters through. In the clearing around the cabin, the sea of green is more uniform than it is farther in the forest, perhaps because of the vigorous stands of sugar maple around the cabin. The red maples, at the edge of the clearing, have a reddish tinge. They, like the ash, have not yet fully leafed out, but they already have flowered and are laden with red seeds that color the whole tree. The spruce tree across the clearing is a dark, almost black green, as are most conifers, although this year's new conifer shoots are bright lime colored, some with a bluish tinge.

As I step out the door, inhaling the cool air, I smell lightness and relief. The birds' calls are intense now, and I see the blue eggshell of a robin on the path—young have hatched. A blue jay flies over the cabin, carrying a twig. Walking on the path, I find tiny violets growing along it: white ones, blue ones, even a few yellow ones. These almost mono-colored miniatures of cultivated pansies are as nature made

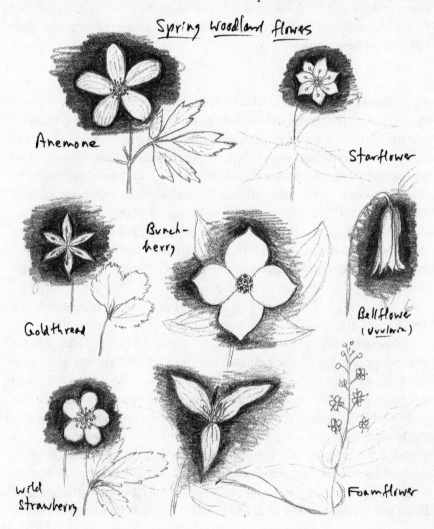

Spring woodland flowers

Anemone

Starflower

Bunch-berry

Goldthread

Bellflowe (Uvularia)

wild strawberry

Foamflower

them, not as someone's whim dictated. Therefore I find them incomparably more beautiful than the domesticated varieties.

Underneath the tree canopy of luminescent green there are flowers everywhere. They are all perennials, and most are snow white: starflower, Canada mayflower, foamflower, goldthread, bunchberry, anemone, strawberry, and Solomon's seal. There is also the white painted trillium with delicate red lines in the center. I had hardly remembered the most obvious and striking spectacle of the myriad snow-white flowers now staring me in the face. It seems always easier to note the one *purple* trillium than all the many white ones.

Jack is soon clamoring for food in ear-shattering, rasping, insistent cries. I feed him, then leave him silent and contented on the branch of a pine by the cabin.

My lookout tree is a red spruce. For the last nine years, I have climbed it to watch for ravens both at dawn and an hour or so before dusk. Its solid branches are stairlike almost down to the ground, because this tree has grown so close to bare ledges that the sunlight has been able to penetrate nearly the full length of its trunk, unlike most of the spruces farther down the slope. However, all of the small scratchy side branches have long been knocked off during my, by now, hundreds of trips up and down.

No brittle-limbed pine tree, this, where strength must be gained through massive thickness. On these spruce branches, barely an inch thick, you can jump up and down and they only bend and bounce back like steel. They've already held tons of snow for perhaps 60 winters.

Red spruce is what forms the "black" tops of these hills and mountains, poking up above the deciduous forests of beechnut, birch, red and sugar maple, and red oak. Mixed in below are also a few basswood and bigtooth aspen, and there are balsam fir, red maple, and white cedar in the swamps. Pin and black cherry, quaking aspen, red maple, and fir trees spring up quickly wherever blowdowns or clearcuts have created temporary openings. All openings in these forests are quite temporary—whenever one occurs, the woods inexorably rush in, like air filling a vacuum.

The snow comes a little earlier and stays a little later up on the spruce ridges than it does below. It's cooler up here, and the thick canopy of needles shades the ground so that the sun cannot melt the snow. This is the breezy year-round home of the tiny golden-crowned kinglet, the black-capped and boreal chickadee, and the red-breasted nuthatch.

The small russet cones of red spruce are borne only near the tip of the crown. Here, under the cover of an umbrella of short, thick green branches laden with cones, I make my perch. I have cleared the live branches to provide a panoramic view in all directions. Here, on top of a hill near Mount Tumbledown, you see an unbroken forest blanketing all the other hills and valleys. Traveling around this wilderness, you see the fresh tracks of moose and deer, as well as bear-claw scratches on the smooth gray trunks of the beechnut trees. You wouldn't guess that the pine forest below my perch was a clear hayfield 40 years ago, surrounded by orchards and sheep pastures.

The old homestead where I built my cabin is visible over there to the north. In springtime, you see a purple lilac that the first settlers planted there; in June, part of the clearing is overgrown with their pink roses as well. The clearing would have been long gone by now, but I keep it roughly trimmed. I like the lilacs and the roses that remind me of the settlers, not to mention the view, and the space in which to grow a few blueberries.

Rhubarb planted over a hundred years ago still comes up near the square granite blocks that used to be a barn foundation. There, among the spirea bushes where I dug down to plant the blueberries, I found the remnants of plows, hay rakes, horse harnesses, and also a square blue bottle with "Hall's" written on one side and "Hair Renewer" on the other.

What happened to the people who lived here? They left their names on the gravestones in tiny cemeteries which are hidden in woodland all around. The weather is already erasing their names from the markers. Of all they built, only the stone walls, crumbling wells, and barn and house foundations still remain as visible reminders.

The gravestones I can still read give only names and dates. The names reflect the names of the local topography—Holman's Ridge, Gleason Mountain, Houghton Brook. The dates reflect a sad truth. Of 100 gravestones that I have read nearby, 57 are for children who died before the age of 20. Mortality between the ages of 20 and 30 remained high (13 gravestones), but then leveled off to four to seven deaths per year until people reached their 90s. Mostly the young and the very young died. The old, once old, endured their lives and missed their children. One headstone for a seven-year-old reads, "Our angel boy." There is no hint on the gravestones of what anyone died from, except one inscription: "Albert M Son of Rodney and Melissa Stearns, a Member of Co G. 17th Regt. Maine Vols., Shot Near Petersburg Va. June 20, 1864." However, I suspect that before antibiotics the same bacteria that wiped out many of the Native Americans also claimed some of the new settlers.

Just below the lookout to the west is a steep granite drop-off with caves. Porcupines live there now, the ones who gnaw scars onto the maple trees.

A few hundred feet down is a tarred road. You can't see it from up here, but you can hear lumber trucks rumble by. Each is filled with two tiers of four-foot logs, with a hydraulic lift resting on top of the load like a big claw.

The trucks go on down the hill toward the paper-mill towns of Jay and Livermore Falls. Others go through Mexico and on to Rumford, not far from here. We're also close to Madrid, China, Norway, Peru, Sweden, Denmark, Naples, Poland, Paris, and Berlin. The rivers that course through this country are the Androscoggin, Kennebec, and Penobscot; the lakes include the Umbagog, Aziscohos, Mooselookmeguntic, and Parmachenee.

The Native Americans who were here when the Whites first arrived belonged to the Abenaki branch of the Algonquin race, and they associated in separate clans and were classified according to the river valley which each group inhabited. Those who lived here in the valley of the Sandy River were likely members of the Kennebec tribe, because the Sandy is a tributary of the Kennebec.

When I was going to the University of Maine in Orono, I lived for a while in a trailer in Old Town, along the Penobscot River. Old Town is the home of the Penobscot Indians where Thoreau hired his guide, Polis, on his way to Mount Katahdin. At a football game I once saw their chief, Chief Bruce Poolaw, dressed in Plains Indian regalia. Poolaw is a Kiowa from Oklahoma.

*   *   *

Not too long after the Mayflower landed on Plymouth Rock, seven Adams brothers came over from England. One of them settled in York, Maine. The others, one of whom led to the line of President John Quincy Adams, settled in Massachusetts. The one who went to Maine had a grandson named Moses, born in Bowdoinham in 1769.

Moses married Martha Kinney of Gardiner, Maine, in 1788. They produced 13 children whose descendants now account for most of the many Adamses in the local phonebooks, plus many more who no longer bear the Adams name. One of the 13 was Mary, who married her father's brother, Jedediah, and thus retained the Adams name. Their son, Jedediah II, was born in 1808 and in turn bore a son, Asa—the Adams for whom Adams Hill was named.

Asa bought the hill farm in 1856 from Orin Perkins, married Elmira Wilkins, and settled there. Their daughter, Flora, was born on the hill in 1858; she married Kendall York and left the farm for several years.

Meanwhile, the hill had passed temporarily out of the Adams family, having been sold by Asa to Frank Morse, and by Morse to Fred Levitt. But Flora's husband, Kendall, bought it back, and the couple returned to live there from 1907 until 1913. By 1913 they already had nine kids.

On Adams Hill, Kendall and Flora raised Ben Davis apples, a variety that keeps well in storage, shipping them to England. The apples were put on hay in the cellar, and remnants of the orchard can still be found in these woods. Beets, turnips, and carrots also were preserved raw, buried in sand, and rhubarb and stringbeans were put up as preserves in jars.

Kendall and Flora also raised sheep and cattle. They had eight to ten cows, which was considered a big herd then. Cream from the milk was brought by wagon twice a week to Weld, the village three miles down the road into the valley, by the shores of Webb Lake. They also sold a few eggs, but had no chicken house. The hens made their nests in the barn or wherever else they could, and you searched for the eggs as necessary. The Adams Hill farm must have been a lot like Floyd and Leona Adamses' in nearby Dryden, where I lived in paradise when I was ten years old, for all of one summer.

After 1913, when Kendall and Flora gave up living here, nobody else lived here either. They sold the farm to their son Richard, who only came during summers, until about 1926. At first the hay continued to be cut. But about 1930, the big barn and the house burned down, and the hay was then no longer cut from the fields. Although the sheep were grazed here for a while afterwards, the black bears that still abound in this area had too great a predilection for mutton, so hefty and feisty Hereford beef cattle were substituted. The cattle were driven up and down the steep hill each spring and fall, well into the early 1940s. Twenty years later, this was a prime deer pasture.

These woods seem alive to me still because the Adamses that lived here were of the same stock as the Adamses that I still know, and have known since childhood. I see these people in the traces of this farm. I see boys herding cows, going fishing, and hoeing corn. I see Floyd, and I see myself.

Thus, Kendall and Flora and their daughter Ethel, who married Alton Adams (a descendant of Martha and Moses), are more than just names to me, because their son Floyd's farm was where I spent many happy times. Leona still visits me on the hill, and I go fishing with their son Bill.

And the connections go deeper. Kendall and Flora's other daughter, Marion, married Ray Ellsworth of Salem, Maine, and it was *their* daughter, Edna Ellsworth, who married Donald "Bucky" Buchanan. Edna and Bucky's son, the "Younger" Bucky, works for the real-estate agency that sold me my land on Adams Hill in 1977. Edna remembers the sour boiled crabapples that her grandmother Flora cooked up on the hill and the rhubarb that was growing next to the then standing barn. The rhubarb, now at the very least 100 years old, still grows exuberantly every spring and still is used for sauces and pies. (I brought both Leona and Edna, who live in nearby towns, a root cutting.)

In about 1958, Edna and Bucky built a small shack next to the ledges up here, calling it Kamp Kaflunk. At that time the fields were still bare. Edna remembers picking blueberries, and being able to see down to Webb Lake. Now trees obscure the view to the lake, and I am harvesting merchantable pine logs where blueberries grew on the open fields only 35 years ago.

When Edna was four years old, in 1917, the brook along the edge of the property was dammed where the footpath now starts up from the tarred road. The aforementioned small establishment called Hildeth's

Mill once stood there and used the water power to make wooden squares for furniture manufacturing. The mill workers stayed at a nearby boardinghouse that was run by Edna's mother. Edna's father ran the saw that processed logs coming out of the mill pond, logs that had floated down Alder Stream from Adams Hill and other hills above. He was, according to those who knew him, "very law-abiding, except when it came to trout fishing."

The local kids went to a tiny one-room schoolhouse in a shack that still stands next to the old mill site. The teacher stayed with the mill workers at the company boardinghouse. The schoolhouse hasn't seen kids since the 1920s, but every year a pair of phoebes raises a brood of four or five young there. The birds come in through a broken window and place their nest of mud and moss on a door lintel just inside. You can also find traces of the dam among the alder bushes; the boarding-house cellar was filled and surrounded by white pine trees that were cut for sawlogs in 1991.

When I was about ten years old, I began to do chores at our neigh-bors' farm, the Potters, mostly milking the cows and filling the woodbox every day. Phil and Myrtle became my best friends. Both worked in the woolen mill in town, and we all used to go hunting and fishing together. Myrtle also made delicious strawberry-rhubarb pie, and as a high school graduation present she knit me the blue and red sweater that I wear still. Myrtle once told me that her ancestors came from the little village of Norridgewock along the Kennebec. She also once told me that her ancestors were Indians, but of course I knew better.

The big lake yonder at the foot of the rounded mountains is Webb Lake. The Webb family still has its farm over on the far north end where the lake becomes shallow and gives way to marshy meadows. Moose often come out of the woods there in June to catch the breeze and escape the blackflies.

Just to the other side of Webb Lake you see a cone-shaped mountain, Mount Blue, less than five miles away. At 3187 feet, it's the tallest mountain around here, except on clear days when you can see all the way to New Hampshire and Mount Washington and the Presidentials, all part of the same White Mountain chain.

A lot of what exists in these woods cannot be seen from my red spruce. Most of the lives around us go on unnoticed. They leave no records. We see only bits and pieces, and then only if we look very, very close, or for very, very long. We have to decipher these other natives of the forest if we want to understand the landscape. A million scents that we never smell waft on the breezes; each of them has special meaning to some insect. If I chop this tree down in the summer, there will be hundreds of beetles of half a dozen species to smell it. They will come flying up against the wind and lay their eggs, which will soon turn into white grubs. Different kinds of wasps will then come to lay their eggs on the grubs, and woodpeckers will later feast on both. There are thousands of beetles of exquisite designs, and not a living person knows even so much as their names.

The birds we know better—at least their names. I've seen 20 species of wood warblers alone around this hill over the years. Each species has its own song, sometimes several. To walk in the woods and not recognize the songs is to not hear them. To not think of the birds' uniquely beautiful and artfully concealed nests is to have the woods seem empty. Most of us are like sleepwalkers here, because we notice so little.

Down there in the valley is a pond. Loons raise their fluffy dark chicks there in the summer, and catch trout. Some springs, they fly in from the winter spent on the open Atlantic even before the ice is out here. If so, they search for other open water and wait to return later. A pair of ravens, too, are familiar with the pond, their home. The pair has nested in the pines along the shores for perhaps 20 years, and it may nest there for perhaps twice as many more.

The indigo bunting that nests in the clearing over there by my cabin spends its winters in Central America, flying the thousands of miles at night, navigating by the star patterns. It learns these patterns as a nestling, huddled down in a little grass cup in the spirea bushes, even as I look at the same sky, so that it can return to its home that I share with it here.

BEGINNINGS

I have a mailbox at the foot of the trail, and rural delivery brings the outside world that far, which is far enough. The newspaper, delivered daily to my mailbox, is a convenience I need to help start my fire in the morning. Not wanting to waste anything, I sometimes even read it.

Ron and Syndi, who live along the road near the mailbox, have a phone line. They also have an outhouse. Knowing I would miss my two kids and my friends, and wanting to hear their voices, I asked, "Hey Ron, can I put a phone in your outhouse?"

"Go right ahead," he said.

The telephone man was incredulous. "You—you want it *where?*"

I added an answering machine, to take incoming calls. Now I have my own private (well, usually) phonebooth.

I don't have a "Tee Vee" to kick in, but sometimes I do turn on the car radio when I drive to Farmington to sit at the diner, drink coffee, and chat with the neighbors. I have a habit of turning it off as soon as I hear the first commercial.

As for the "news," most of what I hear I can do nothing about. This year I want all of my energies and all of my sympathies focused on where they can matter, right here, in the Maine woods.

Jack demands almost constant attention, and, moreover, I need to build an outhouse. I will be caring for and studying wild ravens in an aviary, and I will need to build a foundation under the cabin to prepare it for winter. Wood needs to be cut. My garden needs to be planted. And I want to start running again, all the while accompanied by the flow of nature as a cascade that I stop at leisure to enjoy, not to pass by in the rush of a herd.

I used to be a competitive distance runner, and lately I've started combining my daily jog to and from the mailbox with a run down the road. I go past the lake to see the pair of loons. I keep my eye out for the family of ravens that lives near the edge of the lake. I also pick up small roadkills for Jack, or stop to pick a handful of wild strawberries for myself. While I'm carrying a few berries or a dead bird or chipmunk hidden in one hand, in the other I'm usually carrying a beer bottle or can I've found by the road. I don't know why I pick up these cans. Maybe I'm like a crow picking up shiny objects. In any case, this is excellent training for becoming truly independent, because a lot of people drive by and see me jogging along with a beer can in my hand and I know what they're thinking—that I'm running while under the influence. It's not minding their opinion that's my objective.

June and July turn out to be happy months for me. Many of my journal entries during this time are brief.

May 31–June 2: Settled into camp. Played with Jack. Cut firewood. Many white flowers.

June 3–4: Birdsong survey.

June 5–6: Started to dig latrine. Pouring rain.

June 7–8: Planted buttercup squash and scarlet runner beans. Planted blueberry bushes. Digging trench for cabin foundation.

June 9–17: Clearing trees and brush near seep for possible pond. First fireflies. First deerflies. Sawed wood. Mulched blueberries with the sawdust. Hauled rocks for foundation.

June 15: Watched ravens. Went to town. Planted more blueberries and also asparagus.

June 16: Worked on north foundation wall. Hauled water. Wrote a little.

June 17–18: Cleared path to swimming hole. Hauled more rocks for west foundation wall. Checked out Alder Stream. Big green darner *(Anax junius)* and deep green-blue skimmers *(Libellulidae)* were patrolling it (catching mosquitos?).

June 19: Town day—library, diner, fixed chain saw. Split wood. Checked two graveyards. Went swimming.

June 20–21: Split and piled wood. Blackflies a little less. Finished digging, hauling rocks, and rolling and cajoling them into place—last of north wall built. Was *tired* afterward. Needed nap. Most flowers are done blooming in the woods. It rained buckets. A tiny spider overnight built a most exquisite net over my table. Fed it two mosquitos. It wrapped them up in silk and pulled them up out of net immediately. Watched sap-sucker tree, as usual.

June 22–23: Wood. Worked on the mail.

June 24–25: Many field flowers blooming, especially yellow and orange hawkweed. Did wood. Baked bread, made canoe rack for truck. Long excursions with Jack.

June 26–30: Sapsucker lick drying up? Hawkweed full bloom in clearing. First fireweed is blooming.

July 1 & 2: Chopped wood. Saw paper company forester. Wrote.

July 3: Went to Weld dance. Chopped wood. Wrote.

July 4: Rain. Town to repair chain saw. Phoebe chicks hatching.

July 5: More digging of latrine. Drawing. Walks with Jack. Reading.

July 6: Drawing. Town. Mail.

July 7–9: Trip to the coast.

July 10–13: Fireweed peak bloom. New bumblebee queens are out. Counted 200+ bumblebees of 4 species in a 150-square-foot patch of fireweed. Cut brush. Ran. Drew.

July 14–15: No more swallowtail butterflies. Almost all admirals gone, too. Ran 9 miles.

July 16: Birdsong survey. Cut brush.

July 17–18: Cut brush. Bumblebees tucking themselves into the fireweed blossoms to spend the night. Mostly drones but a few workers. Ran. Most roadkills now are great spangled fritillary butterflies, bumblebees, green snakes, and ground birds (mostly thrushes and yellowthroat warblers).

July 19–21: Finished outhouse! Read.

July 22: Bunchberries are ripe. Still see fireflies. First cricket heard. Ash and sugar maple seeds look ripe. Worked with the ravens in the aviary.

July 23–24: Lots of Virginia ctenuchid moths flying about and foraging on spirea.

July 25–29: Worked with the ravens. Ran.

July 30: See first purple in some of the red maples. Talked with Si Balch, Boise Cascade Co. forester, to evaluate lumbering plan. Blueberries and raspberries are ripe. Found tattered luna moth. Hermit thrush nest w/ eggs—must be second or third clutch.

SCRAP LUMBER

In the Maine woods in early June, along with the birds at dawn you hear chain saws, my own included. I have to work early while it is still cool out, because the blackflies love warmth. When they arise they come by the hundreds or thousands, and leave no patch of skin unbloodied. They hover about in gray clouds that are sometimes so thick you're hesitant to inhale deeply, but a smoky chain saw buzzing at umpteen decibels tends to keep them at bay. I'm "weeding" my sugar maple patch, slashing down red maples right and left, because I'll need the wood for the coming winter, and it requires a few months to dry.

The air is humid. In fact, at this time of day it's 100 percent saturated with water—the glistening dew on the grass attests to that. As the temperature rises, the air takes up more moisture and the dew evaporates. But after an hour of sawing, limbing, and dragging tree stems, I'm bathed in sweat. This abundance of secretions attracts even more blackflies, and they descend as soon as I shut off the saw and pick up the ax for limbing. But I don't begrudge them; they are part of the bargain. It is these tiny critters that help keep Maine green, by keeping people out.

Having finished my chores and my morning bout with the blackflies, up to the level of my tolerance (which is building only gradually), I knock off with a brisk shower under a hanging bucket with nail holes on the bottom.

Refreshed, I retire to the cabin for a second cup of coffee. I now open the windows and let the sun shine in. The mosquitos would swarm in and plague me at night if the windows were left open, but for some strange reason the blackflies that follow me like a plume in the daytime go immediately to the windows, trying to escape.

My old outhouse is dangerously close to topping out. It should be emptied, but I'd rather let the latrine idle for a year or so before I fertilize my sugar maple trees. The solution is to build an alternate outhouse until I empty the first.

Digging a pit in these woods is the hard part. I thought the going would be easy after penetrating the surface layers of soil and roots with spade and ax. But no—down lower, the ground is a solidly compacted mass of clay and rock, which probably is still compressed from the weight of the mile-high glaciers that covered the land until 10,000 years ago. I chip away at the pit a few inches each day with a pick, finally learning why two-seater outhouses were favored here in the old days. You need extra space in which to stand and work, so you end up making a trench rather than a mere pit. And after all that work, you don't want any of that space wasted. Compared to this, building the outhouse above the pit will be a snap.

The hand-painted sign along Route 4 near the town of Jay says "Thick 'n Thin Lumber—Apple Crates, Toboggans." Another sign stuck into the ground near it says "Perot, all new in '92." It's Monday morning, and the wire cable across the road to the lumberyard is down. I drive over it and down the sandy road with aspen and willow bushes on both sides, and I come to a large clearing next to a shallow pond where bullfrogs are garrumphing.

The yard is quiet. An aging German shepherd dog eyes me while she paces back and forth on a long leash, which is fastened near the sawdust pile by a weathered gray saw shed. The dog does not bark. The bullfrogs keep on garrumphing like deep, slow metronomes.

The yard has no neatly piled stacks of lumber. Instead, among some low bushes near the pond, there is a haphazard collection of white pine logs some two to four feet thick, and I wonder how they will ever be muscled onto the sawing block in the old shed. At the end of the shed is a pile of scrap lumber, mostly the facings of pine and hemlock logs with the bark still on; this would do just fine for an outhouse. Opposite the scrap lumber is a long, low building with the word "Office" handwritten on a shingle over the door. That is where I find Parker Kinney this morning.

Parker—the owner, proprietor, and sole operator of Thick 'n Thin— is solidly built and in his late fifties. His gray hair is cropped short and he wears either a blue cap with "Maine" written on the front or a red one saying "Jonsereds" (a brand of chain saw). He has lived in North

Jay all his life, previously having worked as a logger, then at "the Mill" owned by the International Paper Company.

Parker's office is also a workshop filled with apple boxes in various stages of assembly, a toboggan, freshly made bird boxes, a form for bending lathes to make wooden canoes, and another form for making snowshoes. On the walls are small oils of local scenery that he has recently painted. In the far corner is a large woodstove with log benches around it, and a small library of books on the Kennedy assassination.

The last time I came here was six years ago, when Parker sawed the boards for the floors and roof of my log cabin. As he looks up from his work of hammering an apple box together, I notice a flash of recognition cross his face.

"You remember me?" I ask.

"Sure do. How's your pretty wife?"

"She couldn't stand me. Or the woods. Maybe both. She left, but she's doing well."

"You still teachin' in college?"

"Was. Right now I want to build an outhouse. I came to see if I could buy some scrap lumber."

We walk out and he shows me the pile of sidings. "Some of this here will do great," I say.

"Just drive your truck right up then."

On my way back to the office, I pull my chain saw out of my pickup and ask Parker if he knows what's making the chain bind so it won't move. I tell him that the chain sprung off the bar this morning and stuck tight when I put it back on.

"Usually in cases like that, one of the drive teeth has a burr on it. You can usually wear it off, if you can just get the chain to spin once or twice."

He moves the chain forcibly with his hands, then picks up the saw with one hand and yanks the cord to start it with the other. Then he pushes the trigger, and lo and behold, the chain starts up haltingly, then purrs like a Maine coon cat.

"What do I owe you?"

"Nothing. I didn't do anything . . . It was just a matter of knowing."

It may have been "just a matter of knowing" for him, but for me it saves half a day, maybe even the cost of a new chain. I'd gladly hand over $20 and consider it a modest recompense. Knowledge doesn't come cheap in Maine. Parker's gotten his knowledge from years of working in the woods. He knows everything there is to know about chain saws. What I'd be paying for is his past work, if not also past pain (I notice a nasty saw scar across the back of his hand).

"Dammit, Parker, take at least the $10 or you'll make me feel guilty."

"What's it for?"

"Fifty cents for what you did. Nine-fifty for what you knew."

We walk back into the office and sit next to the stove. He lifts open the door to throw in another pine slab. A puff of smoke comes out. "Hot water is here. Coffee and sugar are there. Help yourself."

It is early June, but the mornings are still nippy. The burning wood takes the edge off the morning cold, and it helps brew our coffee. We chat about the lumbering, and the latest antics at Mount Blue, in my neck of the woods.

To protest the proposed cutting of 11,600 cords of wood in Mount Blue State Park, Earth First has reputedly monkey-wrenched machinery and spiked trees. The proposed tree harvest is the result of a deal cut 26 years ago between state officials and Timberlands, a large logging company, for some shore frontage on Webb Lake. What's gotten everyone agitated is the red flag of spiking trees, inflammatory charges that have obscured all other considerations in the controversy.

"As far as I'm concerned," Parker tells me, "they are murderers. Ever see one of those big saws explode? Iron like shrapnel, flying all over the place—saw a man get killed. Got nearly killed myself—a piece flew inches by my face."

Life here in this part of Maine is almost inconceivable without wood, and woods. We burn it for heat. Some cut it for a living. Many earn their livelihood from it by making paper, if not toboggans, snowshoes, apple boxes, or canoes. But it all comes from trees. Trees are our lifeblood, in more ways than one. And that is the problem. There are *woods*, and there is *wood*, and the two have different uses.

THE TIME OF BIRD SONG

Early June is the peak time for bird song, and on June 3, at precisely 8:20 PM, I hear the melody of the hermit thrushes and decide to listen more closely. Official sunset was at 8:17 PM. By 8:25 it is quite dusky, and the hermit thrushes are just coming into stride. But I also hear the following: wood thrush, rose-breasted grosbeak, purple finch, barred owl, and three warblers (Maryland yellowthroat, chestnut-sided, and black-and-white). The bumblebees are still foraging from the chokecherry blossoms in the gloom. A pair of sapsuckers have ringed a white birch next to the cabin, to collect sap and perhaps insects attracted to the sap. Wasps, hummingbirds, moths, and butterflies feed there as well. (I tapped a red maple tree with a chisel, but no sap came. Like the woodpecker, I had only penetrated the bark, removing chunks of it down to the wood. Curiously, although the sapsucker almost never taps maple, it now made tap-holes directly over my fresh chisel marks, as if thinking somebody else knew something it did not.)

Three minutes later, at 8:28 PM, the very first mosquitos venture out into the clearing. It has been a sunny day, and although unimaginable hordes of them are poised and active in the shady woods where the dragonflies don't fly, none has come into the clearing before now, after the dragonflies have retreated. Blackflies are still out, however.

8:30 PM—The mosquitos now come out of the shadows in droves, and their drone almost overwhelms the song of the birds.

8:35 PM—It is dark now, but hermit thrushes, magnolia warblers, and a Maryland yellowthroat sing. The ovenbird, which spends most of its day on the ground singing a loud and monotonous song ("teacher, teacher"), now mysteriously flies high above the treetops, singing a melodious, varied refrain before diving down into the darkness of the forest floor. I've heard it even near midnight on overcast nights, giving a garbled version of this song. This is *some* loony-acting warbler, singing so weirdly in the night, and I suspect there is some fancy evolutionary logic involved here.

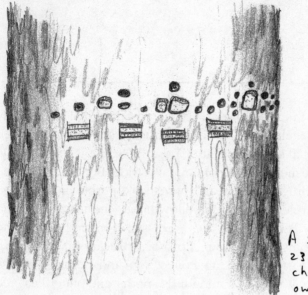

A sapsucker made
23 taps where I
chiseled 4 of my
own — on a red maple

8:40 PM—Now the blackflies are all gone. I hear only the high-pitched hum of hundreds of mosquitos around me, and the soft, clear melodies of a hermit thrush in the distance.

Having heard the evening chorus, I want to hear the dawn chorus as well. And I must do it *now*, because in another week or so, the spectrum of bird songs will not only be different, but will already have begun to fade. From a chart I read that sunrise is at 3:57 AM (eastern standard time). (You need to add one hour to get daylight saving time, which is what my watch is set to.) I want to be up at least an hour before sunrise, so I set my alarm for 3:50 AM.

3:55 AM—The darkness is like night, but there is a faint glow on the eastern horizon. Somewhere from above me a tree swallow is making chirping sounds without pause.

4:03 AM—Stars are still visible in a slightly lightened sky, and I see droplets of dew on the grass. The tree swallow has not changed its cadence. I cannot locate the bird, but I have the impression it is flying continuously high up in the sky. Maybe it took off early, and now it is still too dark to find a place to land? I also hear the first morning call of the ovenbird. The air is cool (58°F). There are so far no mosquitos and no blackflies. However, my scalp, my arms and the backs of my hands are itching more by the minute, as if on fire. I examine myself with a flashlight—my arms are absolutely crawling with a dusting of tiny black pepper-like dots. It's the midges, also called no-see-ums, another of the many biting flies that keep Maine, Maine.

4:09 AM—A mallard calls as it flies in the distance. It's still dark. A barred owl calls down in the swamp. The hermit thrush begins to sing its languid and now so familiar fluting serenade.

4:14 AM—The ovenbird still sings its peripatetic, not yet "normal" song, and the hermit thrush still fifes uninterruptedly. I can now *see* the tree swallow. It indeed is flying above the clearing, as I had suspected, and it still is calling weirdly and monotonously. But five minutes have made a big difference, intermittently I also hear the first later risers: purple finch, rose-breasted grosbeak, white-throated sparrow, and Maryland yellowthroat.

4:25 AM—It is still 32 minutes until official sunrise, but I no longer need a flashlight to write. Now volleys of bird song are erupting from all around, with perhaps ten or more birds singing at once. My spirits lift along with this rising crescendo. I try to pick out the individual species, but it's difficult now. All of the previous ones are singing simultaneously, plus the phoebe, chickadee, junco, and Nashville warbler.

4:35 AM—A rosy salmon sky. At this time a couple of weeks ago, there would've been the spectacular morning display of the wood-cock, but no other birds. Now the exuberance of the varied songsters is almost deafening. I wonder how long through the day the birds will keep it up. Will they still be going full throttle two hours from now, when most people begin to wake up?

4:50–4:55 AM—The volume is already decreasing, and I listen and write down what I hear within five-minute intervals. It is probably still too dark for them to forage. But soon they will probably interrupt their songs as they begin to feed. I hear the swallow (still flying and calling crazily), white-throated sparrow, chestnut-sided warbler, ovenbird, Nashville warbler, and yellowthroat. New birds: raven, red-breasted nuthatch, sapsucker (drumming). The junco and the hermit thrush have stopped.

5:05 AM—It's light! A solitary vireo starts up.

5:07–5:12 AM—The volume is *way* down as only four or five birds sing simultaneously. The ovenbird now sings its standard and familiar "teacher-teacher" song exclusively. Several chestnut-sided warblers are singing full force from the raspberry canes close to the cabin. The red-eyed vireo is singing from the maples near the outhouse (and will continue all day long, with very little interruption). A blue jay screams briefly.

5:30–5:35 AM—The sun is up about five degrees. The no-see-ums are all gone. It is still too cool for the mosquitos, and too early for blackflies. The bird songs are muted; even the swallow is finally reserved. I still hear the rose-breasted grosbeak, sapsucker, chickadee, red-eyed vireo, and warblers (ovenbird, Nashville, and chestnut-sided).

6:15–6:20 AM—The first chain saws buzz, and the first blackflies appear. The same birds are calling as a half hour ago, with a yellow-rumped warbler, hermit thrush, and junco resuming after their respite. Are these now their post-breakfast songs? I take a quick breakfast myself, and a short nap.

8:00–8:05 AM—The sun is burning hot. The mosquitos have already left the field but the blackflies have taken over. They are here in clouds. New birds heard: a winter wren, goldfinches (flying over), and three more warblers (magnolia, Canada, and black-and-white). Others still singing: red-eyed vireo; yellow-rumped, chestnut-sided, and Nashville warblers; tree swallow; ovenbird; rose-breasted grosbeak; phoebe (no song, just contact calls while feeding their young above me as I sit on the front steps).

9:40–9:45 AM—The concert is over. At intervals now I hear only a Maryland yellowthroat, a chestnut-sided warbler, an ovenbird, a swallow, and a red-eyed vireo. A ruby-throated hummingbird flies by. I hear one brief song of the indigo bunting.

3:07–3:12 PM—It is overcast. Birds heard: hermit thrush, Nashville and chestnut-sided warblers, red-eyed vireo, raven.

For the next several days it poured, but I continued to cut my firewood and build an outhouse in the rain. It is pleasant to work in the rain because it knocks down the mosquitos and the blackflies. I also like to hear the rolling thunder and see the flashes of lightning at night, as I lie warm and snug under my roof with the rain drumming over my head but not touching me.

On June 7 it poured all night, and I awoke to a warm, hazy morning. The indigo bunting sang tirelessly from 6 AM until noon. A scarlet tanager came by (and was neither heard nor seen before or after, all summer). In the woods a quarter mile farther down the trail, a parula and a Blackburnian warbler sing every day from morning to evening.

A male bluebird came at about 10 AM, and sang his soft, warbling song after investigating the bird box I had put up by the roasting pit. He sat on a stump, looking around and singing for at least ten minutes. Then he flew off. I wouldn't see another bluebird until the following spring, when a pair came by, again only briefly.

When I repeated the song survey all day on July 16, at the height of summer, the sound was muted by comparison, as mates had been found and most territorial claims made and possibly settled. By then the bird population had likely tripled; most birds had raised their first broods of generally four to five young, and were just starting their second broods. (The young do not sing.)

THE URGENCY OF SPRING

The sun shone, and the phoebe and the robin sang next to our mutual residence, the cabin, almost drowning out the din of the other birds. I drank my coffee and ate boiled oats, then walked down the hill to Alder Stream to pull up a few freshwater clams for Jack. On the way back up, I flushed a hermit thrush off her three pale blue eggs in a nest lined with brown pine needles, sunk into the ground under a foot-tall balsam fir sapling that had new shoots with pea-green needles at the ends of its twigs.

In early June, a flush of tiger swallowtail butterflies has emerged and now sails on warm breezes over the brushy clearing. Red-eyed vireos sing languidly from the deep shady grove of sugar maples, but behind the careless facade of every living thing around me there is a frantic urgency, driven by evolutionary logic, often hidden from most of us and all of them. To them, all this is what Abbey calls "the eternity of the moment," because the future—the winter—comes all too soon.

The fireweed and goldenrod sprouts are already over a foot tall and my measurements show them shooting up as much as half an inch per day, just barely keeping ahead of the grass in the relentless struggle to stay in the sunlight. Meanwhile, the grass is already alive with grass-hopper nymphs and with small green brilliant purple-winged sharp-shooters that suck plant juices and hop erratically.

On the forest floor, the direct sunlight has already been extinguished by the trees' leaves, which are now catching the sun's energy and converting it to phenomenal growth. I don't need a measuring tape to convince me, but I used one nevertheless for a few days, measuring a few specific tree shoots near the cabin every morning and evening. The new sugar maple and white ash shoots have already put on 25 inches, and some of them are still growing at 0.65 and 1.0 inches per day, respectively. At that rate, with two months of summer ahead, they would reach phenomenal height (I once measured an ash shoot sprouting from a cut-off young tree that shot up 10 feet in a single season), but most of the shoots on these trees grow only a little further, before suddenly stopping.

Cut off from direct sunlight, the growth of the spring flowers that decorated the forest floor two to three weeks ago has stopped. In their last efforts to catch the fast-disappearing rays, the late-blooming bunchberry plants even use their flowers as photosynthetic organs. This relative of the dogwood family, whose trees are well known farther south, here creeps along just under the soil. In the spring these underground shoots send up short stems, each with a rosette of six leaves, topped with one four-petaled flower. It takes energy to make the flower and later the red fruit, and all the bunchberry plants growing in the deep shade lack flowers. Those at the edge of the deep shade have green-tinted flowers, while those that spring up where there is more light are pure shining white. In this plant's system, as in most others, reproduction is not subsidized from the outside, and when times are tough there is no reproduction, or the reproductive organs, the flower petals, are used to help make a living.

I wondered how the plant juggles its resources. Would the green flowers turn even greener (and will white flowers turn green) when shaded? To find out, I marked some green flowers and covered some white ones with a black plastic bag. Contrary to what I had predicted, in four days the covered-up white flowers did not turn green, but the green flowers had bleached to white. Apparently only the *young* petals have the capacity to act leaflike. (The next spring, when I observed all the flowers from the beginning, I learned that *all* start out green, and all then turn white. Those in the sun go through the flowering sequence faster.)

In the deep shade of the white pines, small sprigs of blueberry plants barely maintain themselves in the shade. They are *always* without flowers. But the false mayflowers *(Maianthemum)* bloom and do well there. These two-leaved, shade-loving plants form carpets that are now dotted with their white flower spikes. "Shade-loving," however, may be a misnomer. If there were plenty of light, these plants of the shade would be crowded out. They exist here only because conditions are *not* ideal for most flowering plants, and they can get by with less than their competitors.

The ground is covered with a Lilliputian forest—of trees. Down toward the brook, there is a velvety, almost mosslike covering of fresh, inch-long balsam fir seedlings over the bare, shaded ground beneath the mature firs, whose cones large hordes of purple finches, pine siskins and evening grosbeaks fed on last winter. The seedlings grow as densely as four to five per square inch. Up here, in the piney forest next to the hermit thrush's nest where I now find myself, every two or three square-inch area has a small tree on it. Here is also a carpet of two- to three-inch-tall balsam firs, and similarly-sized red maples. In this shade and with this competition, only one in untold thousands of these trees will ever grow taller than a foot. Even now, in a small breeze, the purple seeds of the red maples gyrate down like rapidly rotating helicopter blades, and then carpet the ground.

All of the seedlings, and seeds, lie in waiting. Year-in, year-out, they hardly grow at all. Should a tree fall and a clearing be created, however, a race for the light begins. If the conifers win out, then they shade

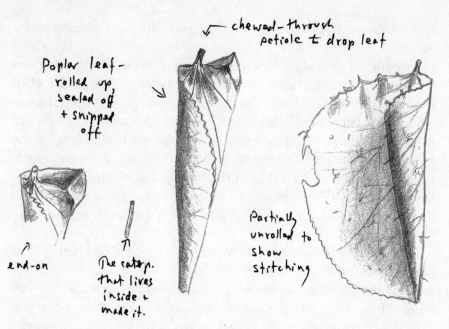

chewed-through
petiole to drop leaf

Poplar leaf-
rolled up,
sealed off
+ snipped
off

end-on

The caterp.
that lives
inside +
made it.

Partially
unrolled to
show
stitching

the ground permanently and only mosses can remain. If deciduous trees gain control, then there is first browse for hare, then deer, then moose. Rich and varied flora and fauna can exist because sunlight still reaches the forest floor for a full month before the tree canopy is finally closed off by leaves in early June. But in unbroken stands of evergreens, the canopy is closed year-round. The herbs on the floor of the deciduous forest must hurry with their flowering before the canopy closes.

The animals, too, adhere to tight schedules. The wood frogs congregated in pools in April. In the space of only three days they called, mated, laid their eggs, and again dispersed far and wide into the forest. Now they remain silent, catch prey, and gain energy to make it through the next winter, which they pass frozen solid under the leaf mold.

Every nook and cranny is full of life, and new crannies are being made. Even as I'm laboriously building an outhouse that will likely become a home for spiders and wood-boring beetle larvae, insects are building shelters and domiciles with ease and grace all by themselves. On one willow tree, hundreds of leaves each have one edge pulled over and held tight to make a tiny purse, the hiding place of a tiny caterpillar who made it. The path to my cabin is littered with poplar leaves that are rolled up and folded and held together with silk, also constructed by a caterpillar. It sits inside and feeds, after snipping its leaf-house off and out of the tree, in order to eat it hidden away from the dangerous bird predators who are hungrily searching above. A little snout beetle does the edible nest even better. It cuts the leaf on each side of the mid-rib, then folds the two sides together, then rolls the double sandwich in from the tip, first depositing a little yellow egg that ends up at the center. The well-wrapped single egg then hatches, and the larva eats its home as its parent leaves. I also have seen many grass stems neatly folded back and forth to create a nest cavity that holds not only a mass of eggs, but also the spider guarding them.

Birds' nests seem clever, but sometimes they seem not nearly as clever as those of spiders and insects that can surely not be credited with a lick of sense.

Everywhere there is a hurry, and little time for dallying, or grieving. Yesterday morning, Jack ate the young cowbird that the phoebes were raising in the nest on the log by my upstairs window. The pair scolded and sounded distraught when they had lost "their" young, but by evening they were singing, and this morning they are already busily refurbishing the nest. In a week they will have laid a new clutch of white eggs, their seeming tragedy having turned to a chance for reproduction.

Last week, the tree swallow pair abandoned its nest box in the old birch tree in front of my window. But by the next day they were flying in and out of another nest box nearby on the pole by the fireplace. Then the male, shiny metallic blue with immaculate white underbelly, sat guard on a dry twig above the box and sang his happy-sounding, gurgling song while his mate, less shiny and more brownish, busied herself carrying long, dry grass stems into their new nest. Soon he brought white feathers. Occasionally she sat still while he hovered above her and briefly touched down to mate.

I checked the old abandoned nest. It contained a clutch of five translucent white eggs. But another female, the mate of only two days before, lay dead in the box as well. There were blood clots on the back of her neck and puncture marks in her breast muscle. I suppose a red squirrel had done it.

The swallows' second nest was highly successful. It fledged four young, and 156 blood-bloated maggots of *Protocalliphora* flies. I would not have known about the latter, except that I saw their squirming white aggregate with translucent skins and red-filled intestines almost covering the entire nest bottom when I opened the bird box for a routine check. The female swallow was in the nest at the time, on top of them, and remained there. I reached in and lifted her out to remove the maggots that were being nourished on her blood and that of her young. For the rest of the summer, she never once failed to swoop noisily within inches of the back of my head whenever she could catch me not looking, but I refrained from swatting her out of the air as I easily could have. To her, I was the enemy. She never once attacked others who came near. That shows amazing discretion, and maybe also shows that while a lot of intelligence might be a good thing, a little can do more harm than good.

On some days, an indigo bunting sings incessantly in my brushy clearing. It's his vocal no-trespassing sign to other males, and it simultaneously serves as a welcome sign to females. Lately I've discovered that on those days when he seems to be absent here, an indigo bunting sings about half a mile away, down along the brush-bordered road.

We humans take time to mourn, and our values condemn highly efficient but ruthless behavior like that of the swallow, and possibly the bunting. But then, we have more time. Most of these adult birds will be dead by next year. All of their lifetime reproduction is often restricted to one year, and in that year it is hazardous, and apportioned to a very specific time. In another week or so, a bird that has lost its mate will no longer find another, even if one might be near, because the appropriate time has passed.

Have *I* waited too long, I wonder? Sitting here on Adams Hill, will I still find my Eve?

June 12
EARLY HARVESTS

Only two weeks have gone by since I arrived, and now the black cherries have started their blooming, and the chokecherries in the clearing have already stopped theirs. The pale pink bell-like blueberry blossoms have dropped, leaving tiny green nubs that will soon become berries. The red maples have just shed their seeds, and two species of fireflies flash their cold white lights on warm evenings. Every morning I awake around 5 AM to the dawn chorus of birds, and by 8 PM I've gone tired to bed, reviewing what I did with the day and what I might do tomorrow.

Winter may seem like a long way off, but it is inevitable. The sooner I get a large stack of dry wood, the better. I have to take out some of the red maples, anyway—the ones that have just about taken over the apple orchard. Most of the approximately 150-year-old apple trees are dead now, having been crowded out by their fast-growing competition. But some apple trees are still hanging on, and without my interference the forest would take them all. I do not want to restore the orchard. But I would like to keep a few of the apple trees alive so that the deer, bear, grouse, porcupine, and red squirrel can feed on their fruit in the fall.

After I've created a tangle of fallen red maples with my chain saw, then comes the hard part of limbing and dragging out the pieces to the "yard" in back of the cabin. The logs have to be sawed into stove-sized lengths. Then they must be split and piled inside the cabin. Although I still have a long ways to go to complete this year's woodpile, the back room is already starting to show results. Every time I glance inside I feel a warm glow of satisfaction. Let it rain, I think; my wood is safely drying. Let the snow and the cold come in the fall and winter; I will *welcome* them. I make the wood-making an ongoing project, to be done for fun, and drag it out for weeks.

This kind of fun is hard on the clothes, and I've worn and torn holes through my pants. I can afford new ones, but find myself spending a couple of quiet hours mending them.

The piles of sawdust from all my work in back of the cabin seemed too good to waste. They make excellent mulching for highbush blueberries, so I planted some. But the ground is rocky. I had to remove boulders to get places to set the plants in. Some of these boulders weigh a good 200 pounds and would make an appreciable contribution to filling the foundation space between the six points on which the cabin rests. I laboriously rolled the first one out of the ground and into its new place in a hole dug under the cabin. Then I thought, why not keep this going? I tried to find still a third one that might fit, then a fourth, and so on until one whole side of the cabin was walled in under the logs. It looked good. Now I wanted to do the other three sides. I mean, there is always a spare moment or two when one needs a break from digging, or sawing. Sometimes rolling a rock into place in a freshly dug hole is just the thing. "Give a man the secure possession of bleak rocks," Arthur Young said in *Travels* in 1787, "and he will turn it into a garden; give him nine years of lease of a garden, and he will convert it to a desert . . . The magic of property turns sand into gold." I love this land and I can't do enough.

JACK

Throughout this time, Jack has been a nearly constant presence. And not an insignificant one. When we first arrived here, on the night of May 30, he was only a few days out of the nest, but he was fully feathered out in shiny garb with a bluish-purple sheen. His eyes were already turning gray from their more youthful bright robin's-egg blue. (In adulthood they turn dark brown). The base of his bill was white, and his mouth, tongue, and palate a bright pink. (All these areas turn coal black in adults.)

On the first day at the cabin, I spent most of my time with him, so he'd know this was home and not just a temporary stopping place. Well able to fly, he was, however, more comfortable walking about. I fed him all the meat he could take, and he took a lot. What he couldn't swallow at the moment, he cached by tucking it into the grass. Unlike older birds, he was not yet covering up this food with nearby leaves and other debris. He spent most of his time alternately sleeping on a logpile by the cabin and playfully yanking at grass, flowers, and leaves.

When I started dragging logs out of the woods, he came and joined me. But after a while he went to sleep in a pine tree near where I was limbing logs. Later he awoke and came when I called him back to the cabin. Jack comes immediately when called if he is hungry, which is most of the time. If he does not come, then he at least answers "krrr . . . krrr . . . ," and I know where he is. He answers me similarly if I'm inside the cabin and he is outside. I reassure him as much as possible that I'm always around, so that he does not wander.

I chose a sleeping place for Jack near the back door on a logpile directly under the north roof, carrying him there perched on my arm. He hopped off onto the logs. I think he liked it there because he quickly went to sleep. He chose the same spot by himself the next evening. I then went there frequently to say goodnight, and when I awoke at night I called him through the logs from inside the cabin. His reply of "krrr . . . krrr . . . ." told me "I hear you. I'm OK." I indeed had a companion. And I was glad.

The next two nights he again chose the same perch under the roof. But then he used a sleeping spot on the dead birch next to my outdoor fireplace, where I linger in the evening. He minds neither the rising smoke nor the drenching rain.

The dead birch also became Jack's favorite singing perch. His song is an almost uninterrupted monologue of gurgling falsetto "quorks," going up and down the scale in pitch, in erratic directions with changes of speed and volume. It sounds like a record or a tape alternately slowed down and speeded up, and he seems to enjoy himself immensely during these sometimes hour-long solos. He usually maintains neutral, flat-feather postures while singing, but sometimes he erects his "ears" and puffs out his throat feathers, as macho adult males do when trying to impress rivals or a potential mate. So it's not just a song, but an act as well. Simultaneously, he keeps busy with his bill: picking, twisting, and pecking at leaves and twigs, or whatever is handy.

Sometimes it pours buckets, and Jack doesn't seem to know enough to get out of the rain. Or he doesn't care. Twice during downpours I put him on the woodpile under the roof, but each time he came back out into the open.

At the end of the first week at the cabin, the white at the base of his bill already had started to get black streaks, and his eyes had turned fully gray. Although by now an excellent flyer, he still showed no interest in using flight as a mode of play. Instead, he amused himself by pulling out the oakum from between the logs of the cabin. He finds irresistible anything that I notice, handle, or appear to think important, so I ignored him, hoping he'd get bored. It worked. In a short while he found new amusement: walking loudly all over the metal roof, to his own vocal accompaniment. He started his roof-walks promptly every morning at 4:30 AM. (To his credit, he sometimes stuffs grass into the same crevices from which he occasionally still pulls out the oakum.)

By June 12, Jack was still begging from me, but even though I'd feed him choice meat and home fries, he'd sometimes spit them up, as if he wanted something else. What might that be?

One afternoon I sat still and watched him for an hour. During that time his motion never once stopped. He took eight consecutive wildly splashing dips in his bird bath (a double-boiler brimful of water), preening till dry in the sunshine after each one. Then he snapped at blackflies flying around his head, yanked up grass and sod, and flew after a swallowtail butterfly. He caught a bee, sang, and pecked at the water bucket.

By mid-June, he'd discovered the joy of flying, and he'd make a wild and exuberant flight around the clearing every morning before I was even up, and then another in the evening before retiring for the night to the top of the bare, open, dead birch. During these flights he'd call raucously, making many fast-repeated, rasping caws. Jack expressed what I was beginning to feel myself.

By another week, flying aerobatics had become his number-one new game. On June 20, for example, he flew at least ten circuits around the clearing, ranging ever farther away over the forest with each one. All of these flights were accompanied by rasping caws, and he slashed the air fast and furiously with his wings, dipping and wheeling and diving. On one flight he spotted a robin, took off after it, and almost closed the distance before it just barely escaped into the woods. I saw him chase butterflies and a yellow-rumped warbler. Once he even took off after a circling turkey vulture. After each flight, he always returned to his birch and preened.

One time while he was preening, I wanted to carry newly split wood into the cabin. He saw the open door, stopped preening, swooped down, and hopped in—totally uninvited. I tossed him out, literally. Barely hitting the ground outside, he hopped right back in, with wing-assisted leaps. Another raven toss. Hey! A new game: he turned around in mid-air and returned all the more quickly. I repeated the toss, to see how fast he'd learn. He learned, apparently, that I'd do it every time, and after 15 times I also learned that I'd tire of the game before he would. So I let him stay in. Now he *ran* all over the cabin in great haste, pecking at the legs of the iron stove, tearing paper, and hopping onto the table. Enough. Out again. This time I closed the door. He responded by flying round and round the cabin, this time not giving his rasping macho calls, but a higher, shorter call that sounded like he might be frustrated.

His raucous morning and evening flights have been picking up in speed, in altitude, and in vocal amplitude and expertise. These play flights give me ideas about my own training as a runner. He never pushes himself to exhaustion. He only exercises when he feels like it and how he feels like it—often and in short, violent spurts.

Jack flies to play. As he flies, even when coming to land on my arm or shoulder, he routinely pulls one wing in, points himself earthward, and does a partial roll during the dive. He seems to be in such high spirits that, after he is fed, he now even does wild jumps that look like he wants to play tag. So I chase him in mock attacks, and he chases me back. He begs, but even after he has fed on the choicest food he still begs, and then rejects what I feed him. Does he want something else that his parents normally provide? Raven young usually follow their parents for a month or so.

"How about a walk to Ron and Syndi's, Jack?" I walk off and he hops along behind. Pretty soon he flies ahead, lands, and walks again. He flushes little white moths. They fly erratically, but he manages to snap them out of the air. I stop to investigate a mushroom. Over he comes, and is diverted by tearing it apart. I walk on. He comes when called. But a half mile down the road after some other diversion, I lose him. On returning to the cabin, I find him already back.

The next day, June 22, was the first time I saw him *cover* his caches.

It rained 2½ inches during the night, and he was out in the open all the while. At 3:45 PM, it was still overcast, after drizzling all morning. However, despite the continuous drenchings, Jack jumped into his bath five times. For the first time, he shook his head at the same time that he shook his wings, to produce a maximal splashing effect.

It is increasingly impossible to split wood. He's started hopping onto the chopping block, grabbing at the ax or at wood I've just chopped, and pecking painfully hard at my legs. My "violence" seems to excite him. Nothing that I touch remains untouched by him and unprobed by his big black bill.

This evening he chased a tiger moth, snapping it out of the air. Then he stood there stupidly when he normally would have swallowed. Instead, he dropped the dead moth and immediately took a drink of water. Tiger moths belong to the Arctiidae, a family of moths that contain noxious, astringent defense secretions, and they are the most beautiful, brightly colored moths around. I'm sure he'll remember the black-yellow-pink color pattern of this species for some time.

On my next excursion to Ron and Syndi's, I jog down the path. He flies along, then lands on my shoulder. I stop and walk. He jumps off, again flushing and catching the small white moths. I jog again, and he flies up to my shoulder again. We repeat this dance numerous times until we're at the bottom of the hill by their cabin under the great pines. Now, suddenly, he flies, circling high over the forest, and is then soon off out of sight. I hear his play-flight sounds in the distance. I call, but there is no answer. Is he lost?

I run back up the hill to the cabin. He is not there. I yell myself hoarse. No answer. No bird. Ten minutes go by. I call again, and then I hear the "krrr . . . krrr . . ." He comes to me over the treetops, lands on top of the cabin, shakes himself, preens, and then comes down onto my shoulder.

Feeling relieved, I sit down on the new stone steps and have a beer. He joins me. I'm surprised how warm his feet are, as well as his bill, right to the tip. At first he busies himself as usual: pecking my shoes and my belt, pulling on my socks, then pecking my leg. It hurts. To distract him, I let him onto my knee and rub the top of his head with my fingers, ruffling his feathers. He now perches quietly, closing his eyes by drawing his bottom lids up. Following my grooming, he does his own: right foot over his wing to scratch the side and back of his head; left foot with same maneuver. A vigorous shake. Then he preens under his right wing, then on top. His other wing, his tail, and then his breast. Another vigorous shake.

I now take him for a walk every day. He seems to greatly enjoy these excursions, and so do I. It's like taking an eager kid out fishing. And like a little kid, he is interested in everything, often getting distracted. If I get out of sight he calls "krrr? krrr?" with an upward, questioning inflection at a fairly high pitch. I answer, "*Here* I am, Jack." He responds with a more even-toned, lower-pitched "krrr . . . krrr . . ." ("OK, I hear you").

On our walk on June 24, it was drizzling. I saw a nest in a pine tree and climbed it to check it out (last year's blue jay nest). Jack perched on a branch and quietly watched. I came down, and we resumed our walk. We came upon several young ovenbirds just out of the nest. He gave chase and quickly caught one and killed it with a snap of his bill. Then he cached it under a rock.

Traveling with Jack allows me to see things from another perspective, and I felt his enthusiasm, despite the rain. It was still raining when we got back. But when I dumped water into his pan, he instantly jumped in for a bath. He'd never dry out today. Waterlogged, he perched on top of the cabin and soon gurgled his song.

He seems to know that he has to keep contact with me when we're in the woods, because as soon as we're separated, he calls me. Yet we spend hours with no contact at all, and he is quite at ease. When I do call him here, he comes instantly, and loudly begs for food, yelling from his perch on my arm into my ear. But when we're on a walk in the woods, he never begs. He is too busy exploring, uninterested in food.

Last week, the Hills Pond raven family of four came by in the clearing. The two young were screaming continually, as if very hungry, and they followed their parents closely. (Jack also does not give me a second's peace, if he is hungry, which is why I keep him well fed.) The two raven parents seemed to move on in a hurry, with the two youngsters always close behind. Yesterday they came again (possibly because they see food in my aviary for my study birds) but this time one of the parents was missing. Had it "escaped" the now almost independent young?

To become independent, ravens have to learn not only about what to eat and where to find it. They also have to learn about the temperament and capabilities of the other predators that will normally kill, tear open, guard, but ultimately provide their daily meat in the wintertime. One way to learn is in play.

On July 2, Jack got a chance to play with a friend's 150-pound white husky dog, now aged and frail. He had met the dog once before, on May 28. Upon first seeing it, Jack had made himself thin and nonthreatening. But after sizing up the old dog, he had changed his strategy. He erected his "ears," made himself broad by spreading his wings, stood tall, erected his feathers to look big and wide, and walked fearlessly towards the jaws of death! The dog turned tail and walked away, which *emboldened* the raven. The dog escaped into the house.

The outcome this time was little different. Jack first yanked the dog's tail, and finding he could get away with that bold gesture, eventually even approached the other end—the one with teeth. First, when the dog wasn't looking, he pecked at a paw. Finding he could

Jack
explores
canid
behavior

easily escape the dog's retaliatory lunge, he next went for the nose. This he never reached, but he had learned what he could risk if such an animal were guarding food, such as a deer carcass. I hoped he would not generalize too much.

On July 5 at 6 PM, I called Jack and he came onto my shoulder after flying in a large circle about the clearing. He made soft sounds of comfort. He was not hungry. Suddenly he froze, standing tall, erect, and sleek. Then he took off in fast flight, and I saw him in vigorous chase after a snowshoe hare. The hare escaped into dense spirea bushes.

Jack's perch was still in my truck, ready for use. And when I was ready to visit my sister Marianne on the Maine coast, I wanted to take him along, remembering our tranquil night journey together from Vermont. It is a long trip to the coast over curvy two-lane roads, through the mill towns of Jay and Livermore Falls, where the clapboard houses crowd close to the road and the dilapidated shops downtown give way to enormous shopping malls on the outskirts. Suddenly, you're on the commercial strip outside of Augusta, where a huge four-lane road passes automobile dealerships of every sort that alternate with every fast-food restaurant known on earth.

The sun was just starting to burn the morning dew off the grass as Jack flew and I jogged down the path to where the truck was parked, next to the stream. I was anxious to hit the road with him at my side. Jack, however, was very hesitant to get in, and he needed a little help. Once inside he hopped about, acting nervous.

We had scarcely pulled onto the tar off the gravel alongside the road when Jack flew off his perch, fluttering against the back window, the side, and the front. His gut geared into action instantly, and for two hours it never let up. I learned that it's impossible to cover with plastic every conceivable place where a frantic raven could leave a mess. He'd find most of the uncovered areas just by random chance. That's the power of energy and persistence, not to mention a well-oiled digestive system. Was that cool feeling down the back of my right arm just a breeze?

Better to concentrate on what *could* be controlled, like the direction he was facing. It was preferable that he face forward and perch on the stick, and I pushed him off every perch except the one I wanted, hoping he'd eventually get the point and learn which position was the most restful. It might work, if restfulness was what he wanted, though I doubted it. To get him to face forward, I turned him by pushing at his tail end. I knew that the more he sat on a certain perch facing in a certain direction, the more he'd do it spontaneously. Maybe I could gradually ingrain the habit. Well, maybe sometime, but not in a few hundred trials, as it turned out.

I pulled over into a McDonald's, and as I was ordering my coffee at the drive-in window, Jack chimed in with a few choice calls. "I *thought* I heard something funny," the girl at the window said incredulously as she handed me my coffee. "That some kind of a haaawk?" I grabbed my coffee to set it in the holder on the dash. Jack grabbed at it too, slashing the styrofoam. Boiling hot coffee shot out in a stream onto my leg.

Later, a moose suddenly trotted ahead of us onto the highway. Jack ignored it. For the most part, he also ignored me. He gave only one kind of sound, a grating, even-toned grunt that lasted about one second. He repeated it thousands of times, like a haywire metronome that had lost its beat. As we slowed down, his noise speeded up, as did his frantic hopping movements, signifying even greater disapproval. He was clearly no longer willing to sit still. Maybe he was becoming independent.

By July 10, some of the wild young ravens were starting to leave their parents (or vice versa). Near noon I looked down from my window and saw *two* young ravens near the front door. One was Jack, but the other could have been his twin. The new one picked up scraps, drank from and bathed in Jack's pan, tried to tear apart a piece of oakum that Jack had previously extracted from the cabin, pulled hair out of a piece of tanned deer hide, pecked at an aluminum pot—in short, it behaved just like Jack. Meanwhile, Jack stayed to the side, ignoring the visitor except for once ambling up behind it, to pull on its tail.

In the afternoon, there were several young ravens without their parents in the woods by the aviary. When I called Jack he begged loudly, and one of these youngsters came instantly to Jack's call, perching in the sugar maple not 30 feet from me. Presumably it knew such begging meant food is being served nearby. Jack kept on begging loudly as I fed him the chewed-up hardtack I was eating myself, and the stranger now looked on. The next day, Jack's begging again attracted other young from afar.

The parental bonds had already loosened. One adult perched nearby with a young, on "Jack's" birch. The young begged only weakly, or the adult ignored it and then flew off down the valley. Although this young followed it, several others that were also in the vicinity stayed.

Jack was not yet independent. He still followed me, which was not always to my liking when I wanted to make a rare trip to the Farmington Diner for a *real* breakfast. So I tossed him a road-kill, and sneaked away on the other side of the cabin. But as I got down by the truck, there was a raven circling overhead. Could that be Jack?

"Jaaack!"

The raven circled down. And just then Ron called: "Hey, how about a cup of coffee?"

"Sure, I'll be right there, if Jack can come, too."

No comment. Eventually I coaxed Jack onto the lawn in their backyard.

We were having coffee and Syndi asked, "Want some pancakes to go with the coffee?"

Ron added, "How about some maple syrup and bacon to go with the pancakes?"

"I never refuse."

Meanwhile, Jack was making himself right at home on the nearby pine tree, where he gurgled on in his monologue song and kept busy picking off twigs and bark.

On my second or third pancake with syrup and bacon to go along with the coffee, it occurred to me that this was an excellent opportunity for an experiment. Jasper, Ron and Syndi's tiger cat, the terror of the birds of Perkins Plantation, was asleep in the house.

"Bring on the cat!" I told Ron.

"You asked for it." He chortled and opened the door.

Jasper hadn't caught a bird, a rabbit, nor even a mouse for at least the last couple of hours, and he hadn't had breakfast either, so now he strolled, unsuspecting, over the "lawn" to our picnic table.

Jack, up in the pine, stopped his idling. He stood up sleek and tall, hunched down, and launched himself for a closer look-see. Cat saw raven coming, crouched, glared with yellow eyes. Jack let out a startled "kraak" and banked up steeply, fluttering to the nearest tree. From there, he studied this strange creature more closely. And vice versa.

After a delay of only a few minutes, he hopped to the ground. Jasper dashed out of sight behind a stump between the two. Maybe the bird would come closer. He did. But Jack was not unaware of what was hiding and where. He just made believe he didn't know. At a safe distance he stopped, forcing the cat to make the first move—a mad dash for Jack, who was having no problems becoming instantly airborne, even while looking over his back.

Jasper soon acted nonchalant again. Jack flew down near him. The scene repeated with minor variations and the game continued to our amusement for the hour that we were having coffee. In between, Jack managed to come by our table occasionally to replenish on pancakes and bacon with maple syrup. But in the end, what he really had gained was having another predator down pat. Having learned that a cat is not a dog, he had made good on an excellent educational opportunity.

"Come on, Jack," I said. "Time to go home."

Three days later, on July 14, I awoke at 5:30 AM hearing his drumming metallic trot and his melodious singing on the roof. Beautiful. Then it was quiet. I called him. No answer. A little while later, I heard a rifle shot down near the cutting. Damn poachers!

Starting damn early this year, I thought. The chain saws then started like angry bees. I heard crashes of falling trees, and the grinding and growling of skidders hauling logs.

At 6:30, I called again. I called myself hoarse. No Jack. This was the first day that he had not been near when I got up, although his exuberant flights lately had looked like he was itchy to go places. I built the fire in my stove, made coffee and cooked cereal, and had breakfast on the front step. No Jack. Every once in a while I called. Nothing.

I thought about the shot I had heard, and contemplated confronting the woodcutters. I started walking in their direction, but turned back. What could I accomplish?

As I was walking back up the hill, I found a four-leaf clover. A bright thought crossed my mind, then this: "Don't be foolish—this is superstition. This might be your first four-leaf clover of the year, but it doesn't mean a thing."

When I got back, I felt pretty low. I made myself another cup of coffee and finished a drawing I had been working on. It was 9:30 AM now. Suddenly I heard his "krrr? krrr?," the calls with the upward, question-like inflection at the end. He was *back*! Overjoyed, I rushed outside and gave him the red-eyed vireo roadkill I'd saved from yesterday's run. Then we went for a walk, with him continually flying on and off my shoulder.

In the afternoon he was gone again. But I was not worried.

In the next few days, he appeared to become a less apt pupil. He'd follow, but suddenly I saw his attention wander, as if his mind were far away. He'd gaze off, then suddenly leave in a wild flight across the forest, as if he were getting impatient with me for not following him. When I gave him a bird or mouse roadkill, he'd fly off with it into the forest, to eat or cache in private. When I scratched his head, he would not hold still as long as before.

On July 19, he left for good. I wished him well, but I also wished he had stayed.

BEDROCK

As I sit atop my spruce tree and look toward Mount Blue to the north, Mount Washington and the Presidentials to the west, and the country road threading like a faint string through the wooded valleys, it is hard to accept the idea that mountains like this come and go. But only about 300 million years ago, during the Carboniferous ages, huge swampy forests of giant lycopods and ferns flourished just west of here. They became the coal beds of Pennsylvania and, together with similar ones in Europe, fueled the Industrial Revolution. This land was near the equator when the coal was formed, and it has been (and still is) a crustal plate drifting over the earth on a sea of molten rock. At the time of the coal swamps, there was still no hint of these mountains, nor was there any Atlantic Ocean.

By the late Carboniferous age, New England had become upland country as the giant Gondwanaland plate slowly but inexorably pushed up from the south. But by the Triassic period, 220 million years ago, the uplands were already being worn down, so that by the Cretaceous period, about 100 million years ago, a broad plain stretched across New England.

Two hundred million years ago, the part of North Africa that is now

Morocco was next to New England, and we were also still located 1,000 miles south of where we are now. As the continental crusts continued to shift, however, they created the rift that would become the Atlantic. The granite that would form the bedrock of these New England hills had solidified from molten pools of minerals deep within the earth. This granite continued to uplift and became Mount Blue, Blueberry Mountain, Mount Bald, and these other gentle hills before me. Once large mountains, now they erode rapidly at about two inches per thousand years. If uplifting ceases, then even Mount Washington, the tallest among them, will be gone in only 30 million years. New England once again will be a level plain.

Mountains come and go, but mayflies now rise out of Alder Stream to dance in the sunlight, just as they did 300 million years ago when they flew out of myriad other streams coming down from mountains that have vanished. There were dragonflies darting after prey in the air then as now, long before the Atlantic Ocean.

Although the lycopods grew up to 100 feet tall and three feet wide at the butt 300 million years ago, we would still have easily recognized them as larger versions of the spore-bearing club-mosses now growing among the moss under the spirea bushes on Adams Hill. Ferns, too, are still common and conspicuous all around me. As I look across the landscape and "see" mountains rising out of the earth, being eroded again to leave plains, and then rising again while the continents drift about, the only permanence I see is in life itself.

In the geologically recent times that are but moments ago, the land here was covered with mile-high ice sheets. The great ice sheets had expanded and retracted rhythmically for the past three million years, and the most recent one began to build up about 100,000 years ago. About 13,000 years ago, it started to melt back during one of the very few intervals of warmth that we are now experiencing. The last glaciers have left so recently that the land is like a house that has just been abandoned, whose new tenants have not yet rearranged the furniture.

The glaciers' tracks are everywhere. Like a bear's claw marks on a beech, you see their grooved scratches on the ledges next to the cabin. Every hillside is a rude jumble of worn, rounded boulders that have been rolled and pushed by ice and haphazardly deposited into giant piles with silt and gravel. Where the rocks were not quite so densely piled, the pioneers attempted to create fields by hauling the stones with their oxen and piling them neatly into long rows or walls.

Steep, winding ridges, or eskers, run north and south. Geologists tell us that the eskers were formed by streams tunneling deep under the huge ice sheets. The streams gradually filled with gravel and left behind these ridges when the glaciers melted.

All around, there are ponds surrounded by arctic bog plants. These ponds were created where blocks of ice left by the retreating glaciers made depressions in the sand. As the ice melted, these "kettle holes" filled with water, and vegetation started to grow in from the edges. The last glaciers have so recently retreated that the earth's crust here has not yet fully rebounded from their great weight, and the Maine coast is still rising. The ocean's water level is rising, too, from more ice melting in the north.

The land at the edge of the ice must have been much like that at the edge of glaciers anywhere today. There was tundra vegetation. There was fireweed with furry yellow-and-black bumblebees foraging from it, and there were bears and caribou. Then, as the edge of the glacier retreated ever further north, the first trees came: the fir and spruce; then the poplar, white pine, and birch. Finally, there came more broad-leaved trees—beech, maple, ash, and basswood—that still are returning to some parts of Maine now.

The caribou have left Maine within recent memory, and extensive spruce-fir forests remain only in northern Maine. The warming trend is continuing, as southern animals such as opossums and cardinals come ever further north, even as the caribou are retreating in the same direction. There is probably not much we can do to reverse Nature's trend. Attempts to reintroduce caribou, in a country now more suitable for deer or moose, have not been all that successful. And only extreme measures, such as clearcutting and herbiciding, allow foresters to grow conifers where hardwoods now want to grow.

At the peak of the Ice Age, when massive amounts of water were locked up in the continental ice sheets, the sea level was low enough to create a landbridge from Asia, between Siberia and Alaska. The woolly mammoth, mastodon, bison, saber-toothed tiger, and other mammals then made their way across to populate America. Humans later followed them to enter a hunting paradise. Within a short time, the camel, rhinoceros, horse, giant beaver, giant ground sloth, musk ox, mastodon, woolly mammoth, and glyptodont all became extinct. It can be safely assumed that, like the moose today, all of these large mammals were tame and unafraid of humans. However, the moose, bison, caribou, deer, and wapiti survived, probably because they were in part or wholly forest dwellers. Yet, even many of these species that remain would not be alive today, for the same reason why all the others died out, were it not for our modern communication system that permitted a coordinated conservation response in the nick of time. The prehistoric Clovis-Folsom hunters who, seeing such plentiful game, killed off most of the large American wildlife, would not have been any more concerned about possible extinctions than were the 19th-century American frontiersmen when confronting the vast bison herds. These animals, too, would have been obliterated, were it not for the accumulation of knowledge that allowed the formulation of a response beyond the local level. In the past, as game got scarce, the hunters merely moved on over the next hill, or beyond the next river. And always, wherever they went, the animals were tame because they had never in their evolutionary history encountered human predators.

Ultimately, as the tundra retreated and the thick, deep woods grew, the only remaining giant herbivores were the forest dwellers living hidden lives. The land was no longer a hunter's paradise, like the open plains of the tundra had been. People lived at the edge of the forest or made clearings in it. But they did not exert brutal force upon the land to try to change it—not, that is, until the Europeans came. Armed with steel axes, with saws, and with a different philosophy, the Europeans not only all but eliminated the animals, but the whole forest as well. Nevertheless, the glaciers' rocky legacy, the harsh climate, and the hordes of biting flies would remain a potent counterforce, and early in the 19th century, when the rich farmlands were opened up in the west,

New Englanders picked up and went in a big exodus. Within the blink of one eye, half a continent had been cleared of forests, and the other half of prairie had turned almost completely to cornfields. Fast-flowing streams of New England then provided the power that turned the wheels of textile and lumber mills. The mill town replaced the farming town, and thus the forest returned. The forest now clothing these hills all around is young.

When Maine attained statehood in 1820, much of the interior was unsettled. The state sold off huge tracts of land and laid them out in six-mile-square "townships," in anticipation of villages being established. But the woods were saved, thanks to the difficulties of farming on land that is a dumping ground of glacial fill. The state's population peaked in 1840, and only passed the one-million mark again in 1980.

Much of the unsettled land, about 10 million acres or about half the state, is "unorganized territory." Today about 17.5 million acres of forest cover about 90 percent of the state, giving Maine the distinction of being the most heavily forested state in the country.

The forest and the land left their impression on the people. My father spent days peering through a microscope at ichneumon flies, and he wanted me to do likewise. But I was stuck at a school while he and my mother left for years to far-off Mexico, then Africa. My heroes were mountain men, who had calloused hands and knew how to handle an ax and a rifle. They were tough men, who did not write books about their exploits, or even talk of them. Some, like Floyd Adams, who had been a Marine in the Pacific, limped and went periodically to the Veterans' Hospital in Togus. They laughed and joked, and sometimes when the topic of the war came up, they became quiet. Mostly, they were not judgmental.

I, too, had wanted to become a mountain man in this boundless wilderness where I presumed no one would ever find me. And as I now looked down at the rushing green water of Alder Stream, I again thought of the flooded stream that had blocked our way as we trudged through the woods to try to live here, 35 years ago, when I ran away from school with my two buddies. It seemed ironic that at long last I *was* here, albeit by a very much more circuitous route than originally planned.

TIME

It's the first week of August. I've been here now for over two months. I came without a schedule and without plans, hoping time would stand still. In a way it has. But that's because every minute of it has been precious. When the moment arrests, then the past and the future evaporate.

But the season rolls on just the same. External time moves inexorably, and the thought of the coming winter intrudes on the "free" time that I crave to observe the detailed lives of plants and animals. There is still much to do before the snow flies. Between a cup of coffee and a serving of eggs, I cannot just pick up the phone and call the oil truck to deliver my winter's heat supply, or the energy required for cooking and lights.

Every time I build a fire to heat up a cup of coffee, it represents precious time that I have had to invest. Time spent choosing a tree; sawing it down; limbing it; sawing it up into chunks; hauling them to the cabin; sawing them up smaller; and splitting, carrying, piling, and feeding the pieces one at a time into the stove that has to be tended by shoveling out the ashes and recycling the tree's minerals back into the woods. I don't mind the time invested for warmth and heat, because the fire also gives more. It flickers and whispers or roars like a live thing that I want to touch or back away from at times because of its fierceness. I tend it carefully, like a companion. It is like attending to Jack was. It's a contract, the fulfillment of which gives structure to my existence. Here and now I need to experience the consequences of my actions, and to exercise the power of daily existence.

To keep warm in the winter ahead, I still have to rechink the cabin and finish the foundation so that the blizzards won't blow through. Chinking involves wedging wads of oakum fibers between the logs, then pounding them in solid with a sharp tool driven by a hammer. I've dug trenches around the cabin, rolled in fieldstones, wedged them in place, carried up 100-pound bags of cement to apply the "mud," and then filled the remainder of the trenches back in with soil. It may not always be fun-for-the-moment. But unlike Sisyphus' boulder, these rocks stay in place.

I'm becoming acutely aware, however, of how much time I spend on simple maintenance. I want to have time to smell the flowers, and end up spending a week just making a place to dispose of my wastes. Even the view toward the mountain costs me hours of too intimate contact with the blackflies, in sweaty work cutting brush, as does having a grassy bank down by the brook. The blueberries that I casually pick are the product of labor digging up bushes in Chesterville Bog and lugging them up the hill with sod attached to the roots, before planting and mulching them with sawdust. Snowshoe hares chew most of them down, raising the worth of those remaining.

Although most of the time I invest seems well worth it and I do it all with pleasure, there is other labor that fritters time away into the dry sand, like water. Therefore I manage it judiciously, trying to cut every corner that I can. Although I do not want to shirk labor, I did not come here just to work.

Back in town, the use of water represents little investment of time beyond the second or so of turning the tap on or off. Here, my water comes from a stone-lined well in the low ground in the woods, some 100 yards below the cabin. I go there and remove a wooden cover that I have made to keep out leaves, lower a bucket down with a rope, draw up the water one bucketful at a time, and carefully pour it through the spout into a five-gallon plastic carboy. Then I lug the carboy back up the hill and set it down next to the cast-iron sink. By using the water sparingly and efficiently, I ordinarily make one water trip last a week. A few cupfuls in the porcelain pan in the sink are sufficient to splash my face in the morning, and the same water is saved for washing my hands later on after heavy work, to then be used once again to soak dishes that are beginning to show encrustations. Dishes only need to be washed once a week, because most of them are not really dirtied after use. My coffee cup, for example, rarely gets washed. Every cupful of heated water is a measure of valuable time.

Although I could easily get along without washing my coffee cup for weeks, gradually beginning to cherish its brown patina finish, I found my attitude toward beer bottles from the recycling center quite another matter. Before I put the home brew in them, I had an obsession to get the dirt out. I washed them in clean fresh water, rinsed them twice, then dried them in the oven of my cast-iron stove.

As for using my back porch as a water- and time-saving device, it is just dandy for some functions, but a damn nuisance for others. The outhouse really was a necessity.

My preoccupation with time, when I want to be timeless, has taken me to new heights of eccentricity. Some friends on their vacation came and we had a wonderful time for two whole days, doing nothing but sitting around chatting on the stone steps of the cabin and on the grassy bank by the stream, casually picking berries, enjoying the view, and cooking meals (each one with newly cleaned utensils, and served on porcelain plates). But I discovered how much I had developed my own patterns, routines, and economies. I got hopelessly impatient. Had I become a hermit already? By the time my friends left, I was a nervous wreck. I accompanied them down to the car, hugged them, said goodbye, and then went on a 15-mile run. After that little bit of unwinding, I dumped a 100-pound bag of Sakrete (a cement) into the metal wheelbarrow I had borrowed from Ron, added water, and mixed it all with a hoe. Then, with bare hands, I packed the cement between the rocks of the foundation of the cabin, felt energized, and did the next bag—and all eight of the others—until the entire project was done before it got dark. That night I slept as if drugged with ambrosia.

August 4
WANDERERS

First the sky darkens. Then you hear low rumbling from a distance. The rumbling becomes louder and is interspersed with flashes of lightning as the sky blackens. The wind picks up, then dies down. A few raindrops patter on the roof. More and more. Now the thunder is drowned out by a roar—the roar of millions of raindrops simultaneously and intermittently pummeling the metal roof. The roar of the raindrops picks up, decreases, increases again, and with it comes the splashing and the crisp crackle of water spattering the ground as it runs in continuous streams off the roof. The streams are thin, but they are many, only one to four inches apart. I rush out to put a barrel under at least one two-foot section of roof to catch some water. In 15 seconds I'm drenched.

The pounding rain continues for about four hours, then subsides to a patter. The rivulets become drips. The thunder subsides. A slight breeze picks up and you see only occasional flashes of distant lightning that still illuminate the whole sky.

By morning thick clouds drift over, but the sky between them is deep blue and occasionally the sun peeks through. As I walk through the meadow in front of the cabin, I brush off the glistening droplets of water adhering to the grass and the spirea bushes. Since leather boots are quickly soaked through in the constant immersion that walking through these fields and brush entails, I have surrendered the idea of dry feet entirely and wear a pair of worn-out running shoes saved especially for this purpose.

To some animals, this moisture is their golden opportunity and perhaps their signal for dispersal. Normally tied to their bogs, ponds, and lakes, they have been released by the rains and driven to wanderlust. During my now almost daily run I saw three run-over newts on the road, all facing away from the pond when they died. Frogs abound. The leopard frogs and bull frogs live year-round in standing water, but recently I've encountered several young leopard frogs in a patch of grass below the cabin. I have no idea where they might have come from, but traveling by hopping, it must have been an enormous distance. As adults, they'll again find ponds.

In *Vermilion Sea: A Naturalist's Journey in Baja, California*, biologist John Janovy, Jr. remarks that humans are among the most footloose of species:

> Pilgrimages seem to be almost instinctive, or at least derived from behaviors now so ingrained in our species that it's difficult to distinguish between genetic and social origins. Of all the animals that migrate, we are surely among the most restless. But humans retain the influence of the geophysical habitat in which they pass their formative years. And often, it seems, we are drawn back to our childhood homes—if not physically, then mentally; if not out of love, then out of curiosity; if not by necessity, then by desire. Through such ramblings we find out who we are.

I am used to seeing wood frogs up here in my forest on the ledge-topped hills. But leopard frogs? It just goes to show how little I know about frogs. And not only frogs; there are also turtles and salamanders. On a steep part of the hill there is a tiny pool fed by a spring. After my run I sometimes stop here to cup my hands and draw up cool water to drink. It is not the place I would expect to see a snapping turtle, but there it was, just below the surface. Did it wander through the forest, up steep mountainsides, stopping here for a refreshing dip? If it had been a large turtle, I would have thought it had been wandering to find a mate or a place to lay its eggs. But no, this one was young and immature. Its carapace measured only four inches.

I had chain-sawed and split a cord or two of wood and left it jumbled in a pile next to the cabin, hoping to stack it after it dried out in the sun. Recurrent rains kept drenching it, and I eventually stacked it wet, deciding that if I waited for it to dry out, I might wait too long.

In the meantime, the woodpile had, like the tiny forest pool, temporarily detained a forest sojourner. Under the bottommost layer of split logs, it lay in a depression dug out of wet sawdust, a beautiful smooth black creature with bright yellow spots. Like the frogs and the turtles, the spotted salamander is also tied to water. It mates, spawns, and grows through its larval stages in forest pools. Now here it was under my woodpile, with miles of forests filled with fallen trees, huge boulders, ravines, and tangles of thick growth separating it from home. How long had it wandered? From where? Why?

The spotted salamander from under the woodpile

The salamander, who now barely moved, could never have moved fast. What compelled it to move at all? I presume it had stopped here to spend the rest of the summer, and the next winter, at this one safe spot it now occupied, but could not have known about in advance. The salamander's tail was turgid with rolls of fat—food stores that would tide it over until next spring, when it might return to the pond from which it came. Or would it find some other pond?

When I consider the journeys of these slow-moving creatures, I am awestruck. Surely, their journeys through strange territory, through a gauntlet of dangers, are not maladaptive aberrations, and our own footlooseness that Jahovy talks about might be more deeply ingrained than we think. We commonly think of evolution working only through selection on the individual, not the population as such, but the genes may manipulate the organism sometimes *against* its own individual interests. The restlessness that moves the turtle and the salamander might more often than not cost them their lives, but the payoff is substantial for the rare wanderer who reaches new shores. Day-to-day survival adds perhaps less to overall future genetic investment than does populating a new habitat. When I build my new pond near the cabin, it will be found by one of those frogs with a footloose tendency. Many hundreds will die in fruitless travels, but the lucky finder will within a couple of years leave thousands of representatives just like itself, and so the genes for dangerous wanderlust will have done well in the lottery of natural selection.

August 6
LATE SUMMER RAMBLINGS

The woods are nearly silent at noon. Hardly any bird can be heard, except occasionally the warble of a red-eyed vireo. At night, too, it is almost silent. How unlike the tropics, where during the day the forest is punctuated by bird song at all times of the year, and at night it always pulsates with insect sounds. Here, neither the katydids nor the crickets have started. But today for the first time I heard the crisp stridulation of a grasshopper. Several days ago I also heard one cicada, but none since.

The butterflies are almost gone. They came as a progression, in discrete species pulses, one after the other. The blues and mourning cloaks flew before the snow or as the snow melted. Then the tiger swallowtails came, and the white admirals after they left, to be followed by the fritillaries and the satyrs, some of which are still active. These are only the visually most common species. Remembering all of them at any one time, it is easy to say, when regarding the moment, "There aren't butterflies like there used to be." Memories often do not specify the week.

The tent caterpillars that had spun their webs all over the apple trees and chokecherry bushes in May have all pupated, and then produced little brown moths. But already the moths are dead or dying. A tattered pale green luna moth lay half-dead on the path. Only the eggs or pupae of these moths now remain. I saw a little glazed brown ring of tent caterpillar moth eggs on a chokecherry twig. It will stay there now until next May, surviving the deep winter cold in an antifreeze solution of sweet glycerol.

On the way around the bend towards Hills Pond on my daily run, I saw skeletonized brown leaves of a spirea plant by the road. They

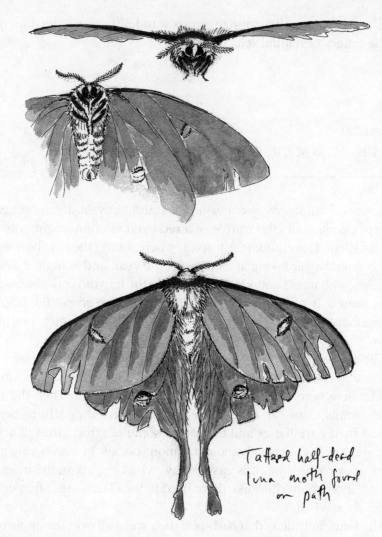

Tattered half-dead
luna moth found
on path

were all covered with a caterpillar web, and many small gray caterpillars were on and in it. I had already seen one of their clusters last week
in my clearing. The caterpillars were all thrashing their posteriors back
and forth, in *synchrony*! As the sun shone on them, the frequency of
their beat increased, and when I shaded them it decreased. Apparently
their rhythm depended on their body temperature, which is to be
expected in cold-blooded animals. But why thrash at all, and why do it
in synchrony? Remembering them, I stopped along the side of the road

to watch this second colony. These caterpillars were *not* thrashing. They were just slowly crawling around. I clapped my hands over them. Nothing happened.

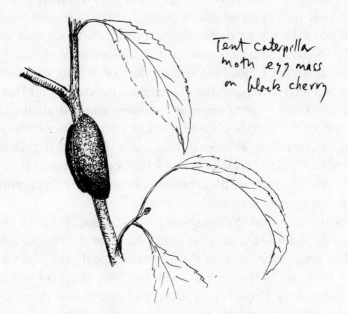

Tent caterpillar moth egg mass on black cherry

I was about ready to leave, as puzzled as before, when an ichneumonid wasp came along. It was a slender black creature with long yellow antennae and yellow-ringed legs. Remembering Papa's collection, I recognized it as an Ophioninae, by its laterally compressed abdomen that expanded from a thin waist barely of pin-diameter to about three millimeters, tipped with a two-millimeter-long ovipositor.

As the wasp hovered near, the caterpillars suddenly started flailing from side to side. The wasp took off as if in fright, but she came right back. Again they started their wild synchronous flailing. Even those that were on a part of the web several inches away and could not have seen her joined in (if these caterpillars can see at all). It was as if an alarm had spread.

As I was bending down trying to observe more closely, a car whizzed by and the draft of air blew through the twigs and sent the wasp off some five to seven feet. But she came right back.

She landed on the web now, then started to approach her quarry on foot. As she came near a caterpillar, she advanced with her abdomen held forward under her, like a dog walking forward but tail first, and then thrust it forward. In that one quick jab she presumably injected an egg through the ovipositor into her victim. The action was so quick that I could not tell whether or not she had connected with the caterpillar that was still thrashing back and forth. But it seemed logical to suppose that the caterpillar's thrashing is a product of evolution, selected because it reduced such parasitism. The egg, if injected, would hatch into a wasp grub that would consume the caterpillar from the inside. The quickness of the wasp's jab is an evolutionary counterstrategy to the caterpillar's thrashing. Presumably there is variation in the response of different caterpillars, so that the quickest ones would still get parasitized, but at a lower rate than their slower neighbors. Evolution works on percentages, not guarantees.

By swinging their abdomens back and forth, the caterpillars keep up not only the defense but also the alarm. And here may lie the answer to why they are synchronous in their movements. If each one moved in perfect antipode to the others, then they would cancel out one another's vibrations. However, by all synchronizing their beating, which they can only accomplish by beating to the rhythm of an already existing vibration, they *amplify* it. The more they amplify it, the more their brothers and sisters (who share their own genes) can be warned.

Even though we can contemplate what the wasp and the caterpillars are doing, and why they are doing it, it is a safe bet that *they* don't. The wasp has no idea that she is laying eggs that will produce other wasps just like her. She does not know why she jabs quickly with her abdomen, or why she jabs at all. The caterpillars do not know why they rock their bodies, much less why they do it in unison. They do not know that their actions have anything to do with a seed that might be injected into them, that will grow into a little white grub inside them and eat them from the inside out, until only a dry hulk of skin remains.

We pride ourselves in knowing what the insects do not know. Yet how many things do we do for the same blind reasons of enhanced survival, all the while just as ignorant of their meaning?

August 9–13
SOME BERRIES

There is a slow drizzly rain today and I try to cheer myself by checking out the nearby Wilton Blueberry Festival. There are no exuberant, wild dancing girls as hoped. In fact, there is not a soul on the streets. I find only an empty tent with empty benches inside. Blueberries can be had, however, at the local supermarket, for $2.79 per quart. Some festival.

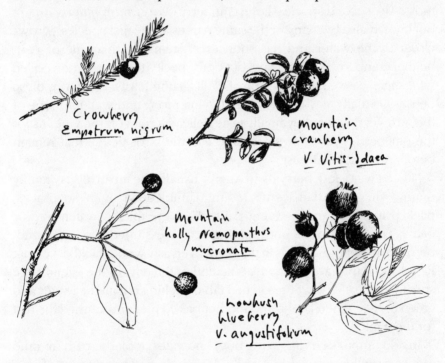

Crowberry
Empetrum nigrum

mountain cranberry
V. Vitis-Idaea

Mountain holly Nemopanthus mucronata

lowbush blueberry
V. angustifolium

Chesterville Bog is only five more miles farther down the road, so I go there. You have to wade among soggy sedges to reach the three- to six-foot-tall blueberry bushes. Their leaves and berries are covered with a film of cool, shiny droplets from the misty rain, but that does not deter a true hunter-gatherer, at least not for exactly one hour by the watch. I picked just a tad under three quarts in that time. You could see where others had been here before, stripping the best bushes. There were two kinds: highbush blueberry *(Vaccinium corymbosum)*, its blue berries covered with a whitish bloom and a slightly acid taste; and highbush huckleberry *(Gaylussacia baccata)*, its nearly black fruit without bloom, sweeter but a bit seedier.

Slightly farther out into the bog, on the green sphagnum moss and along the edge of Pickerel Pond, you now could also pick cranberries. The first species, small cranberry (also called cram-berry, craw-berry, moss-millions, sow-berry, sour berry, fen-berry, marsh-wort, bog-berry, and swamp red-berry, but more accurately *Vaccinium oxy-coccus*), lies like green thread in delicate, slender creeping vines, to which small scale-like leaves are attached, strung over the sphagnum moss. The second species, large cranberry (*Vaccinium macrocarpon*, but known also by many of the same names as the first species), grows closer to the water and has short erect stems. At this time of year, neither cranberry is edible raw, but both species taste fine when boiled with sugar, as we all know from Thanksgiving feasts. However, these berries normally stay on the vine all winter long and well into June of the next year, when they finally taste delicious right off the vine. As for the blueberries ready-sweetened to eat, the birds get all that remain before the end of August.

When I was a kid, berrying in August usually meant an all-day family outing with the Adamses up Mount Tumbledown with knapsacks, packed lunches, and berrying pails. We were told to watch out for bears which might be out berrying too. We stuffed our faces with berries and put the surplus in the big woven ash-wood packbasket that Floyd and Papa carried down at the end of the day. In the evening we spread the berries over the kitchen table, picking out all the leaves and green berries. Then Leona and mother baked pies and canned the rest for the winter.

In addition to berrying in the bog, I now also took the path up onto Mount Tumbledown again. I walked through hardwood forest of very thick sugar maples and yellow birches. Then, rather abruptly, I came into the red spruce belt that is near the top of all the mountains around. But in ten minutes I was through it and onto the glacier-worn bare ledges. Among these granite ledges are veins of vegetation, and small level areas that hold moisture and whose plants are much like those of the floating acid bogs that, through the centuries, have been gradually creeping out into the shallow lakes from the edges. In some cases, they have obliterated the lakes entirely.

The blueberries were ripe. There was the tart blue bog bilberry *(Vaccinium uliginosum)* growing in thickly branching mats. But most abundant was the lowbush blueberry *(Vaccinium angustifolium)* growing in between the boulders. Its berries are light blue, and they look almost white from a covering of bloom that easily rubs off, leaving a deep blue when they are touched. There was still a third species, the low black blueberry *(Vaccinium nigrum)*, whose squat shrubs look like those of the lowbush. The berries have no bloom but are otherwise similar. I pulled up handfuls of the lowbush berries and was quickly satiated. Scores of robins and a raven were also feeding.

Along the cracks in the rocks, I saw short creeping vines of mountain cranberry *(Vaccinium vitis-idaea)* with bunches of bright red berries, some not yet ripe. None are fit to eat raw until after hit by frosts. Also woven in mats up here was the plant known as black crowberry, kinnikinic, or heathberry, among many other names, but more specifically as *Empetrum nigrum.* Its black berries are said to be a favorite of birds in the High Arctic, but they were not very appetizing to me. In the rock basins where larger bog-type plants grew, there was the mountain holly *(Nemopanthus mucronata)* with deep red berries on long slender stems. These fruits taste insipid and contain several big white seeds. Another deciduous holly also growing here on the ledges as well as down in the lower bog was winterberry *(Ilex verticillata)*, which was not yet ripe. But its decorative red berries will feed late-migrating birds, and possibly grouse, who will spread their seeds as they spread so many others. All the berries are designed to be eaten, each catering to the appetite of different animals, or to the same animals at different times. Right now the wintergreen or teaberry has shiny red berries ripened from last year, and this year's white blossoms are beckoning the late-season bumblebees.

A friend from a faraway city was visiting, and she decided to make jam. With the cast-iron kitchen stove already hot from boiling the blueberries, it seemed natural to also put a cobbler into the oven. I made several layers of blueberries and raspberries, each separated by a layer of a

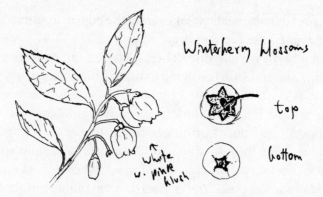

Winterberry blossoms

top

bottom

white w. pink blush

mixture of whole wheat flour with oatmeal, sugar, and butter. That cooking well, with still more room in the oven, why not take some lukewarm water, add yeast and flour, let it rise, and bake that, too? When the bread was done and still so hot we could barely touch it, we broke it open, smelled the yeasty aroma, put butter on thick slices, and had a feast.

The moon was rising, big and yellow above Kinney's Head. It was time for some music—not to create a mood, because the mood was already there, but to give us a beat. It was one of the rare occasions for which I had driven my four-wheel-drive pickup up the steep rocky grade, through the mud, to bring it up to the clearing. We put a tape of Kathy Mattea into the cassette player, turned up the volume, stepped out onto the dewy grass, and in the moonlight next to the fireweed, we laughed and did the Texas two-step, for a perfect Maine blueberry festival celebration after all.

August 14
WHISPERS OF FALL

The fields are daily coming more alive with crickets and grasshoppers, a sure sign that fall is near. It is strange, however, to wake up and not hear a single bird. Sometimes, if you are lucky, you hear a few weak, tentative notes of the red-eyed vireo's song, but he always stops short, as if realizing he's made a mistake. The warblers move about in small flocks, and from up close you hear their occasional little chirps and whispered "tseeps." Otherwise the woods are quiet except for the cicada-like buzz of distant chain saws and the occasional alarm-clock buzzing of a red squirrel.

The colors haven't changed much—the woods are still green. But splashes of red are starting to show here and there on a few scattered red maple trees.

There are many caterpillars. At this time of year I used to hunt for them by searching for their droppings on smooth dirt roads, and by looking for their feeding damage. I later found out that many birds also hunt caterpillars by using their feeding damage as a clue. In turn, many caterpillars have evolved to disguise that feeding damage by getting rid of the evidence: after feeding on a leaf they snip it off at the petiole, then settle down to hide while digesting their meal. On my runs on the back roads I now continually see such fresh, chewed-off, partially eaten leaves and wonder what caterpillar did this and how it is hiding.

The mushrooms are out. In the dark, damp spruce-fir forests, the ground is a brown mat of decaying needles and twigs, dotted with colorful fruiting bodies. They come in brilliant reds, yellows, oranges, greens, purples, pinks, and whites. Caps and more caps, and occasional fingers of flaming orange sprouting out of a log.

The fireweed in front of the cabin is nearly done blooming. Only the tops of the inflorescences are still bearing a few flowers. The bottoms are already shedding seeds. With each breeze, the air is filled with tiny white puffs that drift up and away, as millions of seeds are being launched. There is no end to them. All day long the air is filled with the tiny white specks of fluff that you see only as the sun highlights them against the dark forest opposite the clearing. They rise above the trees, join the sky, and then you don't see them any more. Local events can have wide and unseen consequences.

This morning I did my washing, stopped for breakfast at the diner, got a load of sawed hemlock boards and two-by-fours from Parker Kinney's Thick 'n Thin Lumber, and then ran the 22-mile lake loop. I felt slow all the way, but was never exhausted. Perhaps the cool overcast sky got me down. It seemed so dead—not a bird sang the whole way and the coolness today even suppressed the crickets. I picked my first blackberries on the way, and noted that the chokecherries were ripe but the black cherries still green. Both would soon feed migrant birds on their way down from the north.

When I got back I quickly built a fire to warm up, made coffee to revive, and feasted on fresh bread. Then I split wood, and worked on a sketch before it got dark. No lamp was needed, because I'm usually exhausted by nightfall, when I fall into bed.

My nephew Charlie and my son Stuart are visiting, and we're up at dawn for a trip to Pease Pond.

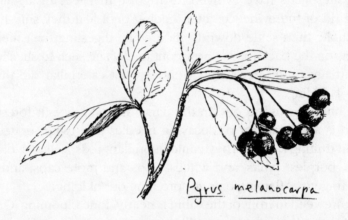

Pyrus melanocarpa

This was the first place I ever caught fish on a hook and line. They were sunfish, beautiful gold-and-green fish with a bright red spot on the edge of each gill cover. A pair of loons with a seemingly very small young patrolled the glassy surface. We paddled down the outlet among the lily pads that were covered with crowds of little black beetle larvae that looked like messy mouse droppings. Then we walked on the floating mat of the bog near the outlet, where the cranberries were still green. A bright pink swamp rose still bloomed near the shore, and the tiny black chokeberries *(Pyrus melanocarpa)* were ripe, but not very tasty. Buttonbush, blue pickerelweed, white water-lilies, and white arrowleaf were also still in bloom. But the red maples at the bog already had a red tinge. Those in the bog are always the very first to show fall colors.

Memories merge with reality now as we beach the canoe near the ledges under the still-standing thick hemlock. Schools of sunfish still loiter there, and Stuart hooks one of their members and stiff-poles it right in. He does not seem as excited as I remember having been at his age. I worry that he has been numbed somehow by too much *Star Trek* at home. But he did get over Ninja Turtles. How much of an artificial, make-believe world will have any meaning for him after he grows up? I still have the fish, then and now, and the anchor to reality forged on long-ago afternoons spent in a place still intact, still moving me.

Charlie plays with a few fish using his fly rod. Then we take a swim along the edge of the floating bog. You have to hunt to see the green cranberries, but the pitcher plants are everywhere—odd, exotic plants found only in floating, perhaps Ice Age bogs like this one.

"Hey, look—it's full of water, just like little cups," Stuart says. He is ready to take a drink from one when he discovers that almost all of them have an insect—a fly, a beetle, or some other partially dissolved arthropod—at the bottom. "How can they eat them?"

"Well, the whole leaf, shaped like a cup with slippery edges, is like a stomach. It secretes juices that digest the trapped insect, and then the plant absorbs it."

The sun turns red through the haze and the water becomes glassy calm. The pond surface is black by the farther shore, over there where we hooked the big catfish at night 40 years ago. As it got dark Billy, Floyd, Robert (a teenager who also worked on the farm), and I used to troll for white perch from our rowboat in the middle of the pond. But today we just paddle across, to again load our canoe and return to the cabin.

My daughter Erica and her friend Robin are also up for the day, and when we get back they have a fire going in the fireplace of loose rocks outside the cabin. We stoke the coals, put on a pot of potatoes, and slap five pork chops onto the grill. Temperatures again have been in the 80s, and the humidity makes the hills look hazy. The blue smoke rises straight up, blending with the haze. Steam rises from the kettle and the pork chops sizzle, licked by flames from the dripping, igniting fat. Against the darkening sky, we see the lumbering flight of large long-horned beetles, which now are droning in noisily, attracted to the scent of the fresh wood we have just sawed and stacked against the side of the cabin. The reedy chirps of the katydids begin to sound, and then it gets dark. We retire into the cabin to eat by candlelight, then come back out to sit around the fire and talk.

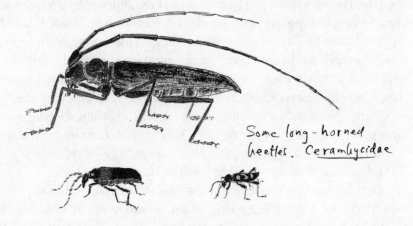

Some long-horned beetles. Cerambycidae

I catch two of the long-horned beetles, admiring their exquisite form. One has antennae much larger than its entire body length. Another of similar size has much shorter antennae. I presume that the length of the antennae is related to the minimum number of sensors needed to detect wood, in the case of the females, and to detect both wood and females, in the case of the males. Therefore, the one with longer antennae probably is the male. Just recently, I have found two other long-horned beetles. One has black and white antennae, and its back is red at the top and black below. The other one is small, wasp-sized. Like a wasp, its back is black with yellow markings. I pin them all, thinking of the other interesting-looking long-horned beetles I saw in June on the mountain ash flowers by the old cabin. I'm being bitten by the collecting bug—I again feel the sensations of discovering and wanting to possess the diverse and the beautiful and the rare, as I had when I was small. Long-horned beetles might be a good group to begin with. They are a group that is not so large as to be unmanageable, for the collector needs some knowledge to keep going. Nevertheless, they are beautiful and varied enough to make one always yearn for yet another. Since there are many species, there is always hope of finding one more. Who knows what this will lead to? It simply feels good to gain even esoteric knowledge, and that is more than enough. Almost immediately after deciding to collect the beetles, I find a slow blue blister beetle that feigns death and exudes noxious yellow secretions from its leg joints when I touch it. The larvae of these beetles wait in flowers to be carried into bees' nests, where they feed on the bees' food stores to complete their development.

Blrstr heetle - soft, dark blve

We slept well. I heard few night sounds, only one short series of the song of the yellow-billed cuckoo. Strange how this songbird is heard mostly at night. I associate it with the alder willow thicket by the old farm. I had not heard it for over 30 years, and now I heard it for only two seconds. Then it was silent.

The summer pleasures as I have known them—the time of bird song, the colorful butterflies, and the excitement of swimming in the stream—have all passed. They seem to have come and gone so *quickly.*

I realize quite palpably now that it is life that has passed, but I feel content that I have lived it. I try to imagine what it would be like if I had the assurance that I'd never die, and wonder if life would be as sweet. The government went so far as to try to eradicate wolves, mountain lions, and eagles—until they were rare. Now it wants to bring them back—*because* they are rare.

August 30

The blackberries grow in dense, impenetrable thickets at old logging sites. Moose and bears walk through here and make paths, but I did not tell Stuart how the paths were made for fear of making him anxious. "We" picked berries for over an hour and collected at least three quarts, but tedium set in after the first ten minutes. Which was all to the good, because it led to the discovery of a coyote dropping full of blackberry seeds.

"Here there's going to be a big blackberry patch," he remarked. He had discovered mutualism. The coyote benefits by eating the berries, and ultimately helps the plants survive.

After we got back, I cooked the berries into jam. Then we planted a remembrance tree. We dug a 5-foot-diameter hole about 25 feet south of the cabin, and then we went down the path to dig and chop out from the roots of neighboring trees a red oak I'd been eyeing for a long time. The tree, about 16 feet tall, 4½ inches in circumference, and with a huge root ball, weighed about 175 pounds. I wheelbarrowed it up, and we pulled off about half the leaves, to reduce water loss. Then we set it into the hole, smoothed earth around it, dumped on a big bucketful of sawdust, put branches on top of the mulch, and then stood back to admire it.

"Someday, Stuart, when that tree is really big, you'll be able to say, 'My dad and I planted that!' "

"Dad, when you die, can I have the cabin?"

"Sure. But let's get this brush cleared."

We cut and piled brush in the field. Then we built our fire in the outdoor fireplace to boil potatoes and heat up red beans for supper.

As it got dark, rather than hear a scary bedtime story, Stuart wanted to walk in the woods with me. Just like my raven Jack, I thought. It had been raining, but the clouds passed and now only the trees continued to drip.

We walked into the pines and sat down on a stone. We listened. Water dripping from the maples among the pines sounded like the walking of animals, because we were thinking of animals.

"Did you hear *that*?" he kept asking. Eventually the breeze stopped and then it became still. Far in the distance, a barred owl hooted. "Who-*whoo*," Stuart called. The owl didn't answer.

We listened some more, and as the darkness closed in Stuart sat on my lap and snuggled closer. We heard a weird, high-toned shriek.

"What's *that*?"

"I don't know. Maybe a coyote."

We had had over an hour's listening time. Over an hour of just sitting still and opening our senses to a strange call in the night, or to whatever stories the woods might want to tell us.

EARLY SEPTEMBER

This is the time of year when the lemon yellow *Colias* and the snow-white *Pieris* butterflies finally flit over the countryside. Where have they been until now? The black mourning cloak butterflies are back again. I last saw them in the early spring while the snow was melting. Then I saw the dense crowds of their spiny black caterpillars on the willows. Now the second generation has emerged in time to overwinter as adults. One was already examining log crevices at the cabin, perhaps looking for a place to hibernate.

The roadkill fauna that I see on my jogs now are the flattened caterpillars of luna moths. These thick, translucent green creatures have dropped off the trees to wander briefly until spinning a cocoon within which the pupa will hibernate. I still see the pale bluish-green, red-studded caterpillars of promethea moths on the ash trees. These, however, are still small. They do not drop off when they reach full size. Instead, they wrap themselves in a leaf on the tree and then spin a cocoon around themselves that they attach to the twig with a strong belt of silk.

The blueberries and raspberries are long gone, but the chokecherries and blackberries are being taken by migrating robins who will soon spread their seeds. The sugar maples are now dispersing millions of their seeds by the wind, and the red squirrels and evening grosbeaks are feasting. The winterberry plants, on a different schedule, are in bloom now. They are of the heath family *(Ericaceae)*, which includes the blueberries and rhododendrons that typically bloom in the early spring.

An ancient apple tree in the clearing has delicious, tart yellow apples of a long-lost variety. It is quickly obvious why they are no longer for sale, even though they are delicious: they are tender and bruise almost to the touch, not qualities suited to today's commercial marketplace.

Yesterday morning I felt slow, and found myself sliding into uncomfortable languor. So I did what I always do to get the engines running again: I gave myself a forced start by going on a run, and after the run I felt better. I mixed bread dough, baked the bread, collected apples, made an apple crisp, sawed a hole into a hollow maple for another tree fort for Stuart, and then we went to the Windsor Fair to catch a greased pig.

It was an hour and a half drive to the fair—enough time to build up anticipation. But when we got there we found out that the pigs were *not* going to be greased, and, moreover, that the contest was no longer open to everyone. Instead, there would be a drawing of ten names for a scramble for 12 *clean* piglets. The pig contest had not only been sanitized, it had also been reduced to a matter of *luck*, maybe to obliterate individual differences in pig-catching ability.

The leveling at the Windsor Fair applied only to humans, however. The animals and veggies still were judged according to the strictest standards of individual achievement. The strength of the horses and oxen was on display, as was the speed and grace of the harness horses. Ordinary-looking tomatoes and cucumbers—they looked exactly like those at the nearest grocery store—were critically judged, and displayed with the appropriate blue or red ribbons signifying their excellence.

We waited at the pulling ring where the anticipated pig scramble was to take place, but the ox pull was still in progress. Great lumbering beasts, yoked together in teams of two, marched out as a pair after a big tractor had pulled the sled with its piled cement blocks back to the starting line. At the command of the teamster, the cattle bent their heads low to the ground, and their great eyeballs rolled till you could see the whites. They pulled and heaved under the prodding and loud yelling of the teamster who tried to coordinate them. Every few yards they stopped for a few seconds' rest and then they were whipped on again. After exactly two minutes the contest was finished, and the tape measure was laid out, the distance recorded, and the tractor again dragged the load back to the starting line for the next pair of contestants.

The horse pull followed, with huge muscular Belgians and Percherons. The weight class we were watching was 3,200 pounds per pair, and the dead weight they were pulling for five minutes was 36 tons.

Each team of horses walked in with heads held high. All had their manes shorn short but their tails were long, though well groomed. These horses did not need to be goaded with a stick. Instead, they were straining at the bits, and the handler had to *restrain* them. These horses were touched by fire—no self-consciousness, no strategic designs, simply unlock and go.

They were only allowed to pull for a few seconds at a time. Had they pulled continuously or slightly out of synchrony with each other, then their combined strength would not have been enough to move their gargantuan load, and *all* of their energy would have been wasted. The point was to go all out for as long as the load moved, and then to stop completely as soon as it slowed down. Thus, by steps, the teams dragged their loads, and although I had not promised myself much excitement watching this spectacle, I found myself strangely drawn in. The horses didn't know how many inches or feet more or less the competing teams had done or would do, and they didn't care. I inwardly cheered the beasts, who were trying so hard that you saw their great muscles bulging, their sides heaving from heavy breathing, sweat pouring off of them, while clods of earth flew high up behind and over their backs as their hooves pounded the earth and all for the pure joy of it. After every pull, the handler affectionately patted their rumps in approval.

Waiting out the time till the pig name-drawing, we took an excursion around to the booths and the rides. I got a cup of coffee, and Stuart, some fried dough with sugar and cinnamon.

Then it was time. Twelve gunnysacks with string had been laid into the now otherwise empty but fenced-in pulling arena. A truck with 12 piglets arrived, and a man with a belly that hung out a foot over his belt and down towards his knees was waiting to release them. People crowded close to the wire.

One after another the contestants' names were called, including Stuart's. I had coached him on the sticky points of greased pigs, and he'd been rehearsing a pig-catching strategy in his mind for days. Only this morning he had shown me a diagram of how he'd intercept a fleeing pig. He had even dragged a log out of the woods, to "practice grabbing hold."

The piglets were released, the announcer shouted, "Go!" and the kids rushed behind them. The group of piglets huddled together in fright as the kids advanced, but in five seconds Stuart and some of the other boys had hurled themselves upon them and were each maneuvering their squealing little porker into a burlap bag. Just like that, it was all over. They came back with bagged pigs to the sidelines, to wait, but three little girls were still chasing the last little piglet who had wised up. The girls waddled behind this animal, who was in no hurry until they came near it. It cockily waited up for them, then deftly sidestepped as the three came near it once again. The piglet made short little satisfied grunts and turned ever sharper corners. It was soon painfully obvious who the winner was in *this* contest. The girls were tiring fast; the pig was not. But here, unlike in nature, failure was not allowed. The announcer called them over and gave each one a $5 bill.

"What are you going to *do* with your pig, Stuart? You *know* your mom won't let you keep it." This reality had not really entered in before, but now that a pig was squealing and wiggling in the burlap bag, this second step in the logical process took over. "You know, I'll bet you could *sell* that pig! I'll bet you could get $10 for it." He could get more, but I had to make him think he was getting a good deal when he saw one.

"Want to see what that man at the ring who brought the pigs might offer? No harm in *asking*, right?"

"Sure."

So we asked. "Give you $20," the man replied, as he reached for his wallet. Stuart's face beamed, and he handed over the pig in the bag.

In the livestock section, a giant of a man next to the goat stalls was selling bunnies. Stuart spotted a dwarf black rabbit in a cage. A little bunny seemed harmless enough. It's cute, doesn't bite, and it's easy to feed and house. It doesn't demand affection to be happy. A bunny might be just the thing for a boy to learn about responsibility for another living being.

"Dad—I *want* one of those. Will you get it for me?"

"But you know you have to take *care* of it?"

"Yes, I'll do everything. I'll build a big cage for it. I'll feed it carrots every day."

"OK."

The man reached in and put it in a cardboard box for us.

"Shadow Black," as Stuart named his bunny, was later installed in a cage at the cabin. Stuart fed it carrots every day. He took it out to hold and to watch it munch clover. He made a sign reading "Shadow Black" and taped it onto the cage, and used a special hole for shoving in carrots. Later we set it free around the cabin. Shadow made its den behind the wood piled next to the cabin, but lasted only one week so near the forest. I heard a scream one night, and then heavy wingbeats.

September 13

Traveling down the long country road by foot is better than driving it because you go slower and see much more. Especially caterpillars. During this season you see many fuzzy arctiids, smooth brown noctuids with red and yellow stripes, big fat sphingids, and plump green silkmoth caterpillars. But because of their sheer abundance, the body count along the highway overwhelmingly consists of grasshoppers and crickets.

I followed moose tracks for about a half mile near Center Hill, examined already-turning red maple leaves from up close, and heard the constant chirping of crickets. I stopped at some blackberry bushes and ate my fill. A bluejay was there, too, and a raven gave its deep rasping territorial calls. A convention of about thirteen blue jays flew over, chattering and making all sorts of weird calls.

Then I heard ravens giving high-pitched calls directly above me and looked up to see a group of eight of them, playing about 2,000 feet up. They were singing and yelling, doing rolls while diving, again soaring up, gradually drifting north. Two split off and went in a southerly direction.

On the purple-blue New England asters along the roadside I saw flies that mimicked hornets, yellow jackets, bees—even bumblebees, and other flies besides. I also saw an ichneumon wasp female creeping down a grass stem, her white-ringed antennae vibrating. She had a yellow scutellum, reddish-brown legs, and a yellow-banded abdomen; otherwise she was black. I had never seen one like her.

As I sat afterwards to let my sweat evaporate, I felt *hunger* and thirst, but also real relaxation and peace. After you've run 20 miles before lunch, the rest of the day is a snap.

September 18

On my daily runs I now notice that even the insect clamor is subsiding. Mostly I hear only the patter of my footsteps and the synchronized swishes of wind over my ears. Occasionally, however, I meet one of my old friends coming down the road to meet me. I hear the swish of his or her steady wingbeats, and the bird cocks its head only slightly toward me. It's one of the ravens from the nest at Hills Pond. I've known the pair since 1984. The birds usually fly 20 or 30 feet above the road, looking for small birds, mice, and other roadkills. I pick up the beer cans and leave them the roadkills. Together we keep our stretch of highway clean.

People ask me, "Why don't you put that energy to some *use*?" But I *am* putting it to use. After a few months, the daily chores of living all tend to grab too much time and make me feel like I'm spinning my wheels. I want to get some traction and *go* places and *see* things.

It is another hot muggy day. In the morning it looks like a smoky haze is hanging over the hills. By afternoon the atmosphere seems translucent blue, like some murky view must have looked through a Silurian sea. Then by dusk it appears as if snowstorms are blotting out the distant views. I get bathed in sweat with the least exertion. I feel hot before I start my runs, but strangely cooler once I go, because finally there is some moving air around my body and some of the sweat can wield its cooling effect.

The no-see-ums love this moist weather. These microscopic biting flies would dry and shrivel in minutes out in the sun. Now they emerge from their hiding places to venture out of the woods and into the clearings. They make the skin crawl like it is on fire, even as it is bathed in sweat. I can outpace them now, but that is only one of the rewards of my runs.

The first bright purple, lavender, and vermilion maple leaves are beginning to sprinkle the path, and I stop to pick some up. Admiring these wonders falling from the trees, I want somehow to preserve them, yet know I have to drink their beauty in *now* because there is nothing that will preserve the fullness of the colors. As soon as a leaf dries, it begins to dull, lacking the luminescence that one full of juices has. I have to see them fresh, every day.

Today I did a 30-mile run, painfully, but at the end I discovered the absolute tip-top of pleasure. The tip-top of pleasure is sitting down at your desk in front of a window with a view toward Mount Bald, with a cup of hot coffee beside you and nothing to do but think or ruminate and scribble with your pen in a notebook.

September 20
CLEARCUTS

On the grass this morning, there was unmistakable white hoar frost. Yet I had to touch it with my fingers to become convinced.

The day became sunny yet stayed cool, and I got the wanderlust to walk in the woods. I need to do it to touch bases, maybe in the same way that I talk with my friends on the telephone, go have coffee at the diner, or read the paper.

I started off going north through the sugar maple grove, admiring the growth of the summer just past and basking in the translucent, shimmering green under their canopy. Ferns were now growing there already, and a few colorful red maple leaves were sprinkled on the old decaying leaves. White Indian pipes were turning their flowers up, and the fertilized pistils were ripening seed. The plant grows only in deep shade.

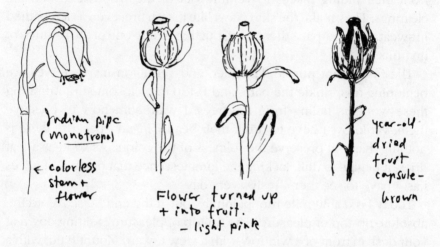

Indian pipe
(monotropa)

← colorless
stem +
flower

Flower turned up
+ into fruit.
– light pink

year-old
dried
fruit
capsule –
brown

Several ichneumon wasps were searching here where sunlight mottled the ferns. They seemed very conspicuous, and very beautiful. But their colors—reds, blues, yellows, blacks, and whites—can only be seen with the eye if you examine the beasts in the hand. I saw their beauty mostly with the mind's eye that had been opened by my father when I was Stuart's age. I was lucky to have been made aware of them.

Most of these wasps were females searching for caterpillars, and they stopped on the ferns only briefly to bask in sunflecks. This year I have already seen nearly a dozen of a big blue-black species with blue wings and bright red abdomen. They are of a group that parasitizes sphinx moth larvae (and only those of specific species) and they are necessarily much rarer than their hosts, which are rare in themselves. For many years I did not see this wasp species at all. Papa, searching for many hours every sunny summer day for 35 years (often with helpers, mostly me), had many species represented in his collection by less than three specimens. Extreme rarity is the rule for many species of ichneumon "flies," which may account for their extreme beauty in my eye.

A flock of blue jays flew over, heading south, but many blue jays stay the entire winter. A winter wren, the small brown stub-tailed gnome of the northern forest, also still remained, and it took a brief interest in me. Making sharp "tick" sounds, it came within four feet of me as I stood immobile. Up and down it jerked in frenetic deep knee bends, then resumed foraging along the ground where it reminded me of an occasionally airborne mouse. For a second it fluffed, then shook itself, vibrating like a bee before flying off. The wren is only here in these woods, I realized, because some large trees had fallen from the wind. The dense brush from the limbs provide it cover, and the upturned roots were its nesting site five months ago.

I go mostly where my feet will carry me, but a rough goal of my excursion today was to climb to my favorite ridge, which until three years ago had been a very rare example of virgin forest—huge sugar maple, ash, basswood, beech, and red oak. Many of the 200-year-old trees were toppling over with age. New ones of all ages were replacing them in progression. It was a climax forest: The big old trees could not get much older nor larger, and the new ones coming in were of the same species. Without some kind of disturbance, this growth pattern and species mix could hold for centuries. These woods had never been touched because, until this century, there had not been bulldozers and skidders that could gain access to this grove tucked between steep rocky slopes and perilous ravines. But three years ago the machines broke through, and after they were done, *nothing* was left standing, not one tree. They left one gigantic tangle of slash that no human could walk through.

As I came to this—to me—almost sacred spot, it already looked considerably altered. The snows of three winters had packed down the slash, and in the summers it had started to decay. Now, as I looked across this giant patch in the forest, I saw a sea of solid green. New growth had swallowed up the slash, fertilizing new growth. I saw fresh moose and deer trails leading from the forest below into this new feeding ground.

The thicket still contained remnants of the raspberry and blackberry patches that always come in right after a clearcut. I feasted on luscious ripe blackberries. However, most of the clearing was now being covered with new trees of my height and taller, so I had a hard time clambering over the remaining horizontal slash and the new vertical shoots that ranged from smaller than finger thickness to one and a half inches in diameter.

I had brought along my camera and measuring tape, and measured out five 100-square-foot plots to count the trees in what I thought were random (or at least representative) locations. But the composition of the trees in each of these plots was still different.

TREE SPECIES COMPOSITION

PLOT	PIN CHERRY	BEECH	STRIPED MAPLE	RED OAK	BIRCH	SUGAR MAPLE	RED MAPLE	QUAKING ASPEN	HORNBEAM	RED SPRUCE
1.	29	16	5	0	0	3	1	0	0	0
2.	4	10	4	3	2	0	13	0	0	0
3.	26	6	15	0	1	0	5	0	0	0
4.	34	5	2	0	1	14	0	1	1	0
5.	13	4	3	0	6	0	0	0	3	1
	106	41	29	3	10	17	19	1	4	1

Glancing over the clearing, you get the impression that the area is being taken over by pin cherry, and my sampling reflects that in most plots. In the hundreds of times I had been here before the cut, I had never once seen a cherry. How and from where had these pin cherries suddenly arrived? If birds had spread their seeds, why were there no choke and black cherries? Both are much more common in the surrounding countryside than pin cherry, though the latter is an especially well-known pioneer species.

The pin cherries will die in less than 20 years. It is difficult to predict which of the trees replacing them will be more "valuable" economically. The species most in demand now might not be the most valuable at harvest time, some 40 years hence, because tastes, technologies, and market needs change faster than that. Species diversity provides economic as well as biological insurance against nature's inherent instability.

I continued on beyond the clearcut and entered a mixed forest with conifers predominating where the ground seemed barren, not green from a cover of ferns and wilting forbs as in the deciduous hardwoods. But here and there the canopy was partially opened by a tree-fall admitting a splatter of sunlight. The ground cover in these places was almost a mat of tiny trees, most no more than an inch or so high.

Even in deep shade, the ground was a Lilliputian carpet of trees so small that I had to get on hands and knees to see them. In one square foot, I counted 14 balsam fir seedlings and 3 red maples, making an average of 17 trees per square foot.

Should a patch of the larger trees be cleared, a terrific struggle would ensue among many, many more competitors than the seedlings that are visible now, because these are only the survivors of the last few years. New seed crops of one or more species rain down almost every single year.

At one place wind had knocked down several small patches of fir trees which now looked, in small scale, much like a clearcut full of slash. Here an impenetrable thicket of seedlings was literally a mossy carpet. Most of this growing mat consisted of balsam fir trees, but there were also birch, beech, ash, maple, and pine. Some of the young pines here showed growth shoots of over one foot during the past year. The firs were growing at half that rate, and the hardwoods were growing even slower, at less than an inch per year. You could see at a glance that this particular patch would be cleared of thousands of hardwood trees by the fast-growing pines that had caught the light. And if the pines were not there, then firs instead would outcompete the hardwood seedlings.

In a clearcut, hardwoods regenerate from *sprouts*, which grow much faster than any seedlings. Since only the hardwoods sprout from young stumps, clearcuts often regenerate to broadleaf (i.e. hardwood) forest faster than to conifers. The practice of clearcutting thus "necessitates" the use of herbicides to foster the growth of conifers that are mainly desired by the paper mills. For this, the government gives the paper companies tax breaks.

I love these woods, even as we harvest lumber from them. But there are different ways of lumbering. There are places in the Maine woods where monster machines snip off whole mature trees, and then also strip them of branches, saw them into sections, and haul them out.

They are let loose by a man in a Plexiglas bubble controlling everything by pulling levers. The monster machines replace hordes of tough
men who left the horse hovel at dawn with their teams of Percherons,
and their chain saws and axes. The men were covered with pitch from
head to foot, and the pitch dried black, and it formed a hard carapace
on their shirts and pants. They sweated and swore and sang, and they
came back to camp in the evening with the satisfaction of a day's work
well done. The monster machines are taking their jobs, and to give
them scope, the lumber companies clearcut huge tracts of old and
diverse forests, replace these with single species of conifers, spew
Agent-Orange-like herbicides from helicopters, and call it forest "management."

AUTUMN

September 23–26
THE GARDEN

The five piles of scarlet runner beans that I had planted together with buttercup squash on June 8 had sprouted but then mostly withered. One bean vine climbed all of two and a half feet up a pole, produced a few scarlet flowers, and left two hanging green pods—just two. "Blooms until frost," the seed package said, "producing long, speckled pods." As for the buttercup squash, the package said, "matures in 90 days." It's been over 100 already. Only one vine grew and now it has one squash, all of three inches in diameter, and about ten flowers. I collected them last night and fried them in a batter of egg yolk and flour, because with the bumblebees gone and even the goldenrods having stopped blooming, I see no future to more farming this year.

In the last couple of days I have hauled in more wood, including a huge black cherry tree that I felled three years ago to get its pinkish wood for stools and chopping blocks. The clearing that the tree created is about 20 feet by 15 feet, and in it there are now vigorous saplings as tall as I am: 30 sugar maples, 3 ash, and 1 red maple are already vying to replace the cherry tree. I also went over the field once more, cutting off the 2 to 10 sprouts that had shot up in the summer, wherever I had cut down the parent tree before. Then I went to do errands in Farmington, stop at the diner, and go on up to the fair, just a mile down the road.

When I came back up at night to the quiet of the hill, I looked for a long time at the mountains. I felt loneliness, or was it a wish to share this beauty and stillness? It didn't seem worth it to cook supper, so I just had a beer and a bowl of corn flakes.

I awoke the next morning to a heavy frost. All the grass in front was brittle and hoary white. When the sun came up, the frost twinkled as if it were studded with millions of diamonds. The leaves on my vine of squash were wilting, and by noon they were dark, limp, and quite dead.

September 27
STARTING A SUGAR SHACK

The stars last night were so brilliant you'd expect them to crackle. But this morning there is a gentle fog that makes the meadow and the woods dreamlike. The sky is lead gray, and so the oranges and yellows of the maples stand out vibrant in contrast. The fog makes everything slightly out of focus, taking the edge off things, so you see only the splashes of color. The moss is luminescent green from the dripping mist, and there is a nutty smell from the decaying vegetation.

I've come back from a 21-mile run along and over hills of a continuous arena of spectacular fall colors. Although the run for some reason left me weak and drained today, I'm now slowly savoring my refill. At the very moment, it's hot coffee with fresh milk and honey.

The first thing I did after I got back was drink a few cans of juice, build a fire, and heat up water for a quick hot shower. I used only about two quarts and let the water gush out over my sweaty head and shoulders and down my flanks onto the grass. It was a delicious sensation, and I savored all two minutes of it.

Now I again have that exquisite relaxed feeling, which I never before had in my life—until I started distance running. I feel the peace and the relaxation that allow for mental concentration. It's different from laziness, which I feel more as a lack of energy after I don't do anything, which is really only a lack of having anything urgent to do. To the contrary, I feel both relaxed and eager—eager to make up what I feel I may have "lost" during two and a half hours on the road.

I needed the long run for reflection and as a punctuation mark after a

wonderful weekend. My daughter Erica drove up Friday night, bringing Stuart. My friend Glenn Booma, a former student, also came. He
brought Bass ale and steak. We built a great fire in the outdoor fireplace
and roasted the steak, drank the beer, and talked. Saturday morning
Glenn went fishing, but first he hauled up water, helped me drag some
more firewood out of the woods, and then carried out trash. He came
back Saturday night without having caught any fish, but having
snagged instead the major ingredients for a fantastic spaghetti feast.

Stuart had forgotten all about his rabbit. He never mentioned it.
Instead, we worked on the sugaring shack for an hour and a half, me
holding the boards in place and he pounding in all the nails. It worked
well—maybe that's why I was so pleased sitting back that evening,
wrestling and play-fighting with Stuart, while Erica and Glenn cooked
spaghetti on the cast-iron stove in the soft golden light of the kerosene
lantern.

October 1

THE FOLIAGE

Canada geese fly over almost every night and every day. They fly by
the hundreds in long orderly V-shaped formations that point south.
The birds call constantly in haunting high cries that make you scan the
skies while they are still far away. I also saw a group of fourteen ravens
fly south. The ravens play individually, in pairs, or in small groups;
they circle high, dive, fold their wings, and shoot up or down with one
or several of their fellows. They chase and frolic, tarry, turn loops; they
make croaks, high cries, and rattling sounds. They do anything but fly
in formation. They remind you of a bunch of schoolboys wandering
down a lonely road, kicking a ball along. The geese fly mechanically,
calling unvaryingly and beating their wings at a steady disciplined
rhythm like soldiers marching off.

There was an inch of ice on the water bucket, and large black and
gray clouds were drifting over from the north. Then the first snow-

flakes descended ominously. But I'm glad: I've made my firewood, maybe three and a half cords. It is sawed, split, and stacked against the back side of the cabin all the way to the roof, against two more sides up to the windows, and in one room inside in four rows up to the ceiling. Let it snow.

The migrant birds have left, except for a few hardy individuals. I saw a flock of five bluebirds, a phoebe, and a hermit thrush. A robin was still feeding on the few remaining black cherries; the chokecherries are long picked off. Several white-throated sparrows still hop in the spirea bushes. On rare occasions one even sings, but haltingly. When the air warms up, you still hear crickets, but the katydids are long dead. Bristly russet caterpillars with a wide black band across the body, the woolly bears, are now crossing the roads seeking hibernation sites and have now become the most common roadkill. According to folklore, the severity of the coming winter can be predicted by the width of the black band, but the width is the same every year while the winters vary.

The ash trees, which are the very last tree to unfurl their leaves in the spring, are now the first to drop them. Half of their leaves are down already. I was surprised, therefore, to find three sphinx caterpillars still feeding on the ash sapling next to the cabin. One of the caterpillars was pea green with light diagonal stripes; the others were yellowish green with brownish purple blotches, blending in beautifully with the now-variable ash-leaf colors. Another species of sphinx caterpillar is similarly camouflaged on a chokecherry bush that still has leaves. Unlike all other, earlier maturing sphinx larvae on trees, these had *not* clipped off their partially eaten leaves. Maybe that is because their fast-dwindling food supply is now not as expendable as it was earlier in the season, and also because there are now fewer predators to notice their feeding tracks. However, although they are evading many birds, they now face another hazard. Since they cannot hibernate, they must complete their growth even while the leaves they feed from and perch on can drop off at any time.

About the only green leaves now near the cabin are those of digitalis (foxglove). The plants sprouted from some pinhead-sized dark seeds that Ernst Mayr had given me years ago. He had received them many years earlier from the late Niko Tinbergen, the Nobel laureate ethologist who would have enjoyed the caterpillars and who would never have guessed that some of his garden is growing on, in the woods on a remote hillside in Maine, tended by an unknown admirer. Digitalis is biennial; they should flower next year. I'm curious to know if bumblebees will visit these, to them, strange flowers. (They did.)

Mostly what's on my mind right now are the fall colors. I go around gaping, as if I have never seen anything like it before. Perhaps I haven't; I'm never quite sure. I could see this display every year and not grow tired of it, like seeing the flight of geese, or hearing the bird songs in spring. I remember, and that might reduce the amazement. But I don't remember the edge—the vividness of the spectacle.

As I scan the forest on Mount Bald that was a rich green only a few days ago, I now see a quiltwork of orange, lavender, gold, purple, lemon, vermilion and cadmium reds, crimson, violet, and peach. The trees luminesce as if emitting rather than reflecting color. With such a varied mix of hues, one would expect a blur, but only at the distance of a couple of miles do they blend to produce an orange-red glow on the hillsides. Even that blend is offset by the nearly black green of the higher, distant spruces. Up closer near the clearing, each tree is distinct: you see a pure, brilliant purple patch next to a pure gold, or red, or brown one.

Peak color is still several days off, but the spectacle is even more pleasing now, because the green that still remains highlights all the more those trees that have already reached their zenith of color. The imagination cannot retain such vibrant, rich colors, and descriptions necessarily fall short.

Looking across the forest or jogging along the road for miles without end, there is hardly a break in the colors, but everywhere you look the mix is different. What trees are responsible for it? I surveyed at least 100 trees of each of the most common species, and was surprised to learn that the key lay in just one species: the red maple. Remove that species, and the whole show would simmer down into a blend of greens, browns, and yellows, with a tinge of orange.

Nearly all of the broadleaf trees up here (seventeen species) change from green to yellow, which tends occasionally toward a golden orange in some species (sugar maple), toward brownish in others (beech), and sometimes toward purplish (red oak, white ash). Even the conifers, the so-called evergreens, are shedding their old needles now (or all of them, in the case of larch), and these too are yellow. Nevertheless, although yellows may *frame* the display, they do not provide its punch.

The maples are the primary show—specifically, the two forest canopy species, sugar *(Acer saccharum)* and red maple *(Acer rubrum)*. Of the 100 sugar maple trees that I examined, four were "orange," four were brownish, and all the rest were golden yellow. The golden crown of a sugar maple tinged with orange can startle you with its luminescence. But it doesn't hold a candle to the red maples, whose color varies from pale lemon yellow to scarlet red to deep purple, with every imaginable gradation in between. Of the 128 red maple trees that I surveyed around my clearing, I rated 33 as "purple," 21 as "deep pink," 52 as "red to orange," and 22 as "clear yellow." There seemed to be no rhyme or reason to the colors. Even the first set of two leaves that tiny seedlings sprout are brilliantly colored, just like those of giant trees. I could determine no correlation with age nor location, at least in the red maples: a purple tree could be right next to a yellow one. In contrast, the few mountain maples whose gold leaves tended toward orange grew in sunny exposed spots, and the few purple- or copper-tinged ash were also the ones more exposed to direct sunshine.

Although from a distance the red maples each looked distinct and uniform in color, that uniformity disappeared on close inspection. The crowns and tips of branches turned color first, so that if you watched a tree in time-lapse, a gradual blush would envelop the tree from the extremities in.

Looking still closer at individual red maple leaves, you saw even more variety. On some trees, the individual leaves were uniformly hued like the whole tree. On others, however, it looked like a mad spray-painter had been busy: yellow leaves might be marked with bold red blotches, or finely speckled in red or pink.

The leaves of the red maple drop at the height of their color, and all the while that the forest is ablaze in color from the underbrush up through the tips of the crowns, the ground also is aflame as the magic settles onto the wilting ferns and last year's decaying brown leaves. I want to pick up every leaf, for each one seems brilliant and unique. I know that the colors are even more precious because they are ephemeral—in a few days they all fade to a uniform brown.

To remind myself of what they look like, because my memory will fade almost as quickly as the leaves themselves, I picked up those that struck my fancy as I walked up the half-mile path from my jog down on the highway. From that short walk I compiled the following list, in testimony to the astounding variety of color, to be browsed at random in the same fashion as one might enjoy the leaves themselves:

yellow with small purple blotches

green center with shining red
   margins

light lemon yellow

yellow with dot-sized red speckles

uniform bright vermilion red

uniform pink

yellow with washes of red and
   orange

greenish yellow with one bright red
   corner

orange with large red blotches

bright uniform luminescent red

pale yellow with green veins

green with red spots

yellow with red washes and
   mottling along veins

greenish yellow with diffuse red
   washes

red with purple veins

yellow with purple edges

gold evenly speckled with red

yellow with green mottling

red with yellow veins

pale yellow (almost white)

gold with three bright red dots

purple with green blotches

bright vermilion red with yellow
   veins

uniform peach

uniform orange

peach grading into pinks and
   yellow

yellow with dots that look like
   blood

It seemed odd that color should vary radically even within a given leaf. Why should two leaves on the same tree, with genetically uniform cells, have totally different colors? Was I looking at a *progression* of colors that all leaves underwent, uniformly—say, from green to yellow to spotted to uniform red to purple? To find out, I marked 20 individual leaves with tags attached by dental floss and rechecked them when color was complete and they started to fall off. I learned that a leaf that *started* to turn yellow continued until it was fully yellow, then dropped off. The same process occurred with a red or purple one. Any spots showed themselves early, and they did not enlarge, contract, appear anew, or disappear. It was as if each green leaf destined to have spots had a spot pattern program within it that predetermined *one* specific pattern. On the other hand, the overall pattern or base color cannot be immutable from the time the leaf first forms; because when a branch is partially broken or damaged, the leaves on the broken part turn a different color than the rest. The later the break, the less the difference in color. In the same way, leaves that are picked and allowed to dry are arrested in their color development—they dry green, green-yellow, green-red, or whatever color they had when they were picked.

In the previous year when I had thinned out the red maples from my sugar maple grove, I had already been impressed with the various autumn colors of the leaves from the red maple sprouts that had regrown by that fall. I had photographed them on individually identified trees, hoping to compare the colors with next (this) year's. Of the 15 small trees I revisited, all had similar colors as before: previously purple trees tended to be purple again, yellows again yellow, and so forth. There *was* some uniformity to the fall colors after all.

October 3
THE ANNUAL RENDEZVOUS

A few years ago I invited everyone I knew up to the hill at the height of the foliage season to celebrate. We celebrated. We also decided it would be a good idea to celebrate again. So we came back the next year, and the next. Now this first weekend in October has become a tradition.

Bill Adams and his son Cutter, in Mohawk haircut, came at 10 AM, bringing the 68-pound lamb for the roast. I put it on my shoulders and carried it all the way up without setting it down once, trying to prove to myself that my back had recovered from an old automobile accident. We wired the meat onto a long pole of green red maple, got a fire going in the firepit, and soon had the lamb on the pole suspended over red coals between two stakes driven into the ground.

People drifted up the hill all day long, gathering around the firepit and the beer, where we were still congregated long after dark. Bruce, my running teammate from UMO, came from Rhode Island. He is still dreaming of running the good marathon. Wolfe and his wife Denise came all the way from Boston. I think he is the only man I ever hugged. He is a big bear of a fellow who was at the original rendezvous. Alice, who came from New Jersey, tells me he said I'm the only one whose funeral he'll go to. I know why. I had previously instructed him to do it right—told him to spread the ashes round a maple tree, and to draw all the money he needed from my account to get kegs of beer and have a blast, preferably with naked dancing around the campfire. Jack Daniels was there, too, escorted by Delia in remembrance of our former raven-cage-raising and raven-catching parties.

I love the company and I have vivid dreams as I see others appreciating the beauty of this land, too. I'd like Stuart to grow up here, feeling solidness, security, companionship, and rootedness of place. I "see" myself with my Eve, a beautiful earth-mother woman, making a home here. We'd have two others here—we'd work until sweaty and dog-tired, clearing woods, raising sheep and honeybees, making maple syrup, tending trout ponds and berry patches . . . We'd carve out a homestead.

I can feel a common flow of energy suffusing all of us as we crowd around the campfire. We're all "Brothers and Sisters of the Raven." What's scary is that it all sounds so idyllic—so *un*real in relation to the "reality" that we live—that we tend to reject it almost automatically. But what if it is really possible?

Stuart was wound up all day in the excitement. But at night, he wanted quiet time with me—to lie down in the forest alone and listen. We saw a woodcock fly by on rapidly beating wings, like a giant nocturnal hummingbird. We heard a barred owl call, and he answered it and it answered back.

October 6
CHERRY HILL DRIVE

The trees are disrobing in earnest now. Their leaves fall in droves when the sun hits them in the morning, and then they settle onto white hoar frost. Fog hangs like steam over the pond in the cool morning air, but the sun later clears it away, leaving the loons swimming serenely over the glassy surface.

I drove past Hills Pond, then up the logging road to Cherry Hill. Twenty years ago this road was overgrowing with alders, but you could still squeeze through. At that time, after about a mile, the road still continued on through the woods to the small village of Carthage, but instead we took a sharp left and went up a steep rise to a little plateau with a clearing of hayfields, where aspen were creeping in from the edges. I loved that spot. I thought it was the most beautiful spot in the whole world. You had views toward all the mountains. Mount Bald's slopes, clad in hardwoods and topped with red spruce, were just to the right as you stood between the collapsing barn of gray weather-worn boards and the abandoned small frame house with a cellar and two dormers. I had imagined children growing up here in solitude at the foot of the mountain where the moose, deer, and bear roamed. To the north they would have looked toward Mount Blue, and to the west lay the great Webb Lake with Mount Tumbledown and Mount Jackson beyond.

There is now no longer any sign of the farm except for the stone-lined cellar hole and old split-granite barn foundations. The forest has reclaimed the fields even in this short time since I was last there. There are no more views.

Once traversed by horse-drawn buggies and sleighs, the road that had served other farms beyond was later used mainly in the fall by hunters. But the old apple orchard that once attracted deer has now, like many other small old apple orchards, been taken over by a forest of ash, red and sugar maples, birch, and aspen.

Now the road serves pickup trucks, logging trucks, and skidders (large-wheeled, all-terrain vehicles used for pulling logs out of the woods). These woods have been heavily logged in recent years, and they are still being logged. About two miles in, I came to a lumbering yard with many freshly dragged-out logs and a skidder. On the way up the hill, I had noticed the stumps of trees cut in previous years. The area had been, one might say, intensively harvested. It was not, how-ever, "clearcut," where there is a deliberate attempt to cut down every standing stem in order to herbicide (treeicide?), in order to plant something alien, such as a spruce plantation in maple territory. Instead, here the loggers had come in and simply cut the biggest trees, leaving the rest. The light created from this selective cutting had been exploited by all the young trees standing in the shadows, and after a mere ten years there was now a forest that did not have trees as large as the original stand, but did have innumerably more of them.

Coyote and bear scat, or droppings, were on the rough rocky gravel of the road. There were also fresh deer and moose tracks. The maples and aspens along the roadsides were vibrant with color. The opened habitat looked prime for grouse, woodcock, and small forest birds. Dead and dying trees had been left to rot, as had the slash that was now decaying into the ground. In a mere 50 years, an expert would be unable to tell the difference between this forest and one that had been left alone. If we must harvest wood, and I think we must, then this sloppy, messy way of going about it without "management" seemed like not such a bad way to go. Certainly in Maine.

October 7
HOUSEMATES

As winter is approaching, many creatures are moving into the cabin. For the last couple of weeks, I've been especially irritated by a crowd of mice. They are all deer mice—cute little creatures with big black eyes, clothed in velvety fur, either gray or light brown on the back and white on the belly. I was quite willing to share my groceries with them downstairs, but things are getting out of hand with noise upstairs. It is their caterwauling and running around at night in the ceiling directly over my bed, in energetic crowds, that eventually makes them less than appealing. Even one of them can become annoying when you know that they whelp a dozen pups at a time at regular biweekly intervals.

One by one I caught the mice in the six traps I set every night. The nightly bedlam gradually subsided. It was peaceful last night; maybe I've got the last one.

In the daytime there is another problem. Although I chinked the cabin tightly (I thought), lately I've seen crowds of cluster flies. Cluster flies are bristly gray insects, about five times bigger than the ordinary housefly, *Musca domestica*. They are harmless enough, because they are generally outdoor creatures. Their only known sin is that they parasitize earthworms: they lay their eggs in living worms, and the larvae then consume the worms from the inside out. But in the winter, cluster flies aggregate in thick black crowds jammed into cracks and crannies inside the house. This is well and good, except that whenever the temperature gets a degree or two above freezing, they consider it springtime. And when cluster flies awaken in crowds from a sound sleep, they buzz crazily to the nearest window, trying to get out. Outside temperatures may be -30°F, but they don't know that. In past winters, I have opened the window saying, "Go for it!" In a big black cloud they then all rush out, and are stopped short in a second or so as they freeze up and nosedive into the frigid snow. The voracious shrews and chickadees await them. I close the window again with a smug smile on my face, and wait for the next battalion of eager buzzers.

It is not cold yet. But the flies have anticipated the winter nevertheless, and they are already well ensconced in the cracks of the cabin. I notice a few on the windows now and then, wondering how they could have breached my thorough chinking that I hope will exclude the cold, if not them.

It is warm today. In fact, it is downright comfortable inside the cabin after I've built a fire to make my coffee and cook my oatmeal. I hear a buzzing—there on the upstairs windows, they have gathered. I estimate there are over 1,000 flies. Each one of them is vigorously popping against a windowpane. When one comes crashing down after a lot of buzzing, it picks itself up after only a short pause, to rub its wings by reaching over its back with one or another of four hindlegs. Then it rubs the front and hindlegs together, as if removing bothersome cobwebs. Then, after rubbing its great bulging eyes with the frontlegs, it is ready to give it another go at the unyielding pane.

October 8
MOOSE

"Moose" was the last word on Thoreau's lips when he died, and moose seem to be the first thing on everyone's mind these days in Maine. The newspaper just ran a picture of a big bull taking a dip in a swimming pool in Lewiston, one of the biggest towns in the state. Almost every day there is a picture on the "sports" page of a smiling person with a rifle, standing over a dead moose.

The six-day moose hunt ends in two days. Nine hundred Maine residents and 100 nonresidents had their names drawn from a lottery. The success ratio of those who draw and shoot a moose is usually over 90 percent, and most hunters are choosy: they want bulls. Three bulls are taken for every cow. The herd is still increasing, thanks to Maine's massive clearcutting that has produced prime moose pasture. Next year even more names likely will be drawn from the lottery. I did not put my name in. First of all, I don't need hundreds of pounds of meat. Second, I'm a deer hunter. Although I don't hunt for the "sport" of it, shooting an animal that is big, tame, majestic, and stupid—like a moose—is not my idea of fun. It's a little like taking meat off the grocery shelf, with a lot more work added. Strangely, I have never seen pictures of smiling persons with shopping carts standing over piles of steak.

I have tasted moose meat, though. I picked up a fresh moose head out of a barrel at Castonguay's livestock butchering place in Livermore Falls, where I get meat scraps for my ravens. It was that of a bull, I presume, because the top of the skull had been sawed off to lift the set of antlers and on its chin hung the long skin dewlap. The brain lay revealed. For the command and control center for a nearly half-ton animal, a moose's brain is not large—no bigger than one of my fists. I was surprised that the tongue was still intact. In the old days, when Thoreau came to hike and canoe in the Maine woods, the tongue was considered the best part. Moose (and bison) were killed just for the tongue and then the rest was left. Now people take the rest and leave the tongue. I'm not biased either way, so I boiled the tongue and had it for a supper, or two or three. It had a smooth, soft texture and tasted fine. I put the rest of the head out onto the ledges on the top of the hill for the ravens. When I checked at dawn one morning, I found ravens perched all around, "talking it up." And talk is all they did at first. Always first the talking, then the eating. They seemed to like it fine, too, once they got over their initial inhibitions.

Hoping to see a live moose, I've lately been avoiding the beer can alley, the tarred road, and have been running on deserted logging roads. My favorite run is along a dirt track that leaves the highway near Hills Pond and in about a mile comes to two summer camps at a bridge over Alder Stream. Nobody lives beyond that, and the track then crosses a temporary bridge of poplar logs lashed together. It continues as a logging road that was used seven years ago. The topsoil that had so recently been bulldozed already has layers of green hairycap moss. Everywhere my feet brush through the now dried stems of Canada cat's-foot (*Antennaria canadensis*), Canada and flat-topped gold-enrod (*Solidago canadensis* and *Euthamia graminifolia*), through various species of sedges, asters, and fleabane (*Erigeron*). There are wild strawberry plants in the old wheel tracks of the logging trucks, with birch seedlings sprouting among them. Here, in the summer, there would also be tunnels of solitary bees nesting in the sun-warmed soil.

Just beyond the bridge, the road follows the brook filled with large black rocks and gravel. The grade is uphill. A forest of spruce and balsam firs, mixed with poplars, crowds to the edge of the road so that your view is limited to some 10 or 15 feet into the dark interior of the woods. But within a mile I come into an old cutting, which extends for many more miles, almost to Mount Blue.

This cutting was technically a clearcut. Seven years ago it must have looked like hell, with the road plowed by a bulldozer, skidder trails leading off in all directions, and a continuous slash field of devastation, with only old or broken trees left standing.

Here Nature had reasserted herself, and an objective appraisal not biased by what *was*, gave a very favorable if not a grand impression. The hills had now come into view, and I enjoyed the grand spectacle of Mount Blue ahead. I saw various ridges to my right and left, undulating over the huge and imposing landscape that now was exposed. It was an arena of fast-living new trees engaged in a wild, fierce free-for-all of competition for the sun's energy, hogged for decades by a previous generation of large trees.

Right now the pin cherries and striped maple were ahead in this competition, as is typical. There were thickets of them, and they were already over 15 feet tall in some places. The lanceolate cherry leaves shone deep crimson against the luminescent lemon yellow of the broad striped-maple leaves. And the quiltwork of intense color stretched as far as the eye could see. The white birch, beech, and red and sugar maples were not far behind in the race for light. This was now moose country.

Beneath this almost heaving, leavening growth of new forest, lay the old rotting slash. Each winter the snow had come, fastened itself upon these twigs, and pressed them down. Each summer, the ferns had grown through and up over the twigs. Each autumn, the leaves had fallen on the pressed-down twigs and the wilting ferns upon them, providing even more purchase to the next year's snow. Underneath them now was a haven for shrews, for red-backed voles, and (in places) for snowshoe hare. The rodents multiplied, along with hosts of insects. Soon they attracted predators—weasels, foxes, fishers. The rotting slash served as fertilizer. It also aided growth by holding off the direct summer sunlight to keep the soil moist, and by partially insulating the ground from the intense winter cold. Slash might not be pretty to look at, but like manure, it is a rich and useful resource.

Now, seven years later, what the loggers had left imparted a look of wilderness to the area. The wind had taken some of the standing trees. Others had died. Some had flushed out in new growth. Everywhere there were dead, broken gray snags of old balsam firs whose flaking, peeling bark would provide nest sites for brown creepers. There were dark green spruce sentinels, ideal for black-throated green and myrtle warblers. There were jagged sugar maples, dead or coming back to life, where nuthatches and hairy woodpeckers would nest. It was a wilderness in flux, with overlapping diversity from young to ancient trees, one that would be as different seven years from now, as it had been seven years ago.

The twitch trails (for "twitching" or pulling out logs) to the sides of the dirt road I was jogging on were now an impassable tangle of blackberry and raspberry vines that were already being replaced by birch and maple seedlings. Now, instead of trucks, the road was used by moose, bear, coyote, and deer. The moose had left the most obvious impressions: their large hoofprints were sunk into the earth, while here and there young trees were mangled and broken with their bark scraped and their twigs left dangling loosely. It is not clear why bull moose attack trees and clumps of bushes, but I guess if you are a 1,000-pound bull moose, you don't need a logical reason to do whatever you feel like.

The further along I went, the more the road turned into what looked like a well-traveled moose trail. Passing depressions full of water off to the side, I saw hoofprints in the puddles, whose bottoms still had stirred-up mud. I expected to see a moose any minute. And, as if on cue, I did.

He was walking down the road toward me, as I was jogging toward him. I stopped. He stopped. We stared at each other, some 100 yards apart. I now casually started walking toward him. He did the same, now making occasional soft whining grunts. When we were about 20 yards apart, it became obvious that this half-ton beast was not going to turn aside for a mere creature such as me. I increased my speed toward him, but only to hastily reach a small maple tree next to the road between us. I climbed without delay. He stood and watched.

After I was well situated about 15 feet up, I too watched. The moose resumed his leisurely walk toward me, and his grunts. As he got closer I could hear his breathing as well, and I saw his pink tongue sweep across his lips. I saw the whites of his black eyes as he rolled them up at me. They were bloodshot.

When he was just alongside me he again stopped and briefly contemplated this strange creature in blue T-shirt and jogging shorts up in the sugar maple sapling. I counted 13 tines on the left side and 12 on the right side of his majestic palmate rack. One of the tines had been broken recently, possibly in an encounter with another bull, or perhaps with a tree. I broke off a twig and tossed it down. It hit his antlers and bounced onto the ground. He leisurely looked down at it, then up again at me. The black fur on the top of his back, possibly seven or eight feet off the ground, started to rise at the ridge of his back as he looked me over. He rolled his eyeball once more, then slowly lifted his long white-clad legs one by one, and ambled on, reminding me of the Pleistocene epoch. About 50 feet past me, he casually turned off to the left, down another overgrown logging road.

Dusk was coming now. The sky was salmon. I heard croaking and glanced up to see six ravens in three pairs flying high overhead toward the east.

I slid down the tree, wiped loose bark and lichens off my shirt, and felt much refreshed and perhaps a touch exuberant. I continued down the road, and on the way home stopped briefly at Ron and Syndi's to tell them about the moose. That called for a beer or two, and a supper of fresh corn, while sitting around the fire outside. Ron said, "I wonder what the rich folks are doing tonight."

It was long past sundown when I finally continued on back to the cabin, trotting up the path in the moonlight.

All the trees looked bigger. I saw the dappling of the moonlight on the different layers of leaves above, and those leaves already fallen onto the ground seemed silvery.

October 10
HOME CHORES

In the night I woke to a tremendous blowing, followed by thunderous pounding of rain on the roof. With so much blowing and so much rain,

I somehow expected the cabin to be flooded, so I got up and checked around with a flashlight. But all was shipshape, and I slept all the sounder despite the pounding.

This morning the rain has stopped. Fog hangs on the distant and some of the near hills. The overnight wind had ripped many leaves from the red maples, and the color of the ground carpet beneath them is continuous and intricate. It is a random arrangement of all the colors imaginable, with continually different color combinations at every glance. Those leaves that have fallen bottom-side up are glistening with water dimpled into tiny pearls, while those that fell top-side up are uniformly moist.

The fog rising from the distant red and yellow hills looks like it could be a huge forest fire, savage and lustful. Instead, because I know it is cool dampness on rising mists through brilliant foliage, it looks very soothing and friendly. The scene, too, changes like a forest fire, but in slow motion. The maple that was red and purple is totally bare now, just across the clearing. The sugar maple beside it that was green is now a shining golden yellow, touched with orange over the outer branches as if colored with a hot breath.

I stay inside this morning, doing the chores that I hate. Over the summer I have developed routines that reduce these chores from what elaborate exercises they could be. There has been no conscious thought or plan. As happens in organic evolution, it's just how things work out by always following the path of least resistance with the available material.

I've done very well without refrigeration, and at the same time saved enormously in washing dishes. The solution is my big red kettle. I can both cook and eat out of it, and it seldom needs to be washed. The red enameled kettle always sits on the back of the stove. Whenever I have meat or vegetables left over, in they go. Do I feel like having potatoes? Good—I throw some in. Peas? Likewise. I might add a little oregano, garlic, onions, salt, and butter once in a while. But basically, the kettle is my reverse refrigerator. Once or twice a day it gets brought to a boil, and then all the bacteria are killed. The lid is left on tight, to reduce new invasions. Thus, unlike a refrigerator used to retard bacteria by cold, I have a kettle to kill them by heat.

Breakfast is of an entirely different flavor. As soon as I get up, I light the fire and put on a stainless steel pot for hot water. In ten minutes, it is boiling. Part of the water goes into my coffee; the rest stays in the pot, to which I add oatmeal or some other grain. In five minutes it is cooked, and I eat it out of the pot. Then I put in a little water on the bottom, and put it back on the stove to soften. Next morning I can clean it out with a swirl or two of clean water. Presto—I'm ready for the next breakfast. I use one spoon to eat the soup, another to stir the coffee and then the cereal.

There is no excuse for store-bought bread, when it is so easy to make your own. The cookbooks treat bread-making like some kind of a voodoo art, and each person tries to outdo the other in more "creative" recipes. I get a headache just looking at a cookbook. There are lists of ingredients telling you how many teaspoons of this, how many pounds of that, how to mix it, at what temperature to hold it, how to shape it, what to put it in, how long to let it rise, at what temperature to have the oven, and how long to bake it. Just for starters they expect you to own measuring cups, weighing scales, thermometers, clocks, and a damn good ability to follow directions. I don't have any of these accoutrements or attributes, but I still make a passable bread.

The trouble is that when there are so many directions, you lose sight of the direction. You are paralyzed by the thought that you can totally flub up by missing a step. For example, if you heat the dough to 108°F rather than 104°F, might you end up with rolls as hard as rocks, and as tasteless? The directions are not very helpful because you don't know which are the relevant variables; you don't know whether the shape of the pan is more important than the amount of yeast you put in. So you have to follow every step slavishly. Actually, neither the amount of yeast you use, nor the shape of the pan, makes any real difference. If the directions would merely explain *how* bread is made, so that you could understand the process, then you could do it and never ever buy another loaf.

Making bread may be science—the art of mastering detail while keeping a distant vision in mind—but it is the opposite from trying to understand and manipulate wild nature. When making bread, you try to control only *one* kind of organism. In an ecosystem like a forest (as opposed to a spruce plantation) you have, literally, millions. And the closest thing to a law of nature that I know of is that those who try to run an ecosystem inevitably get the opposite results of those they intend. That's because it's just too darn complex to understand.

Here is my "recipe" for making bread: Bread dough is a culture of yeast cells. They need a little salt for their minerals, flour or sugar as an energy source, and warm temperature (heat kills them, and cold immobilizes them). When the cells grow they give off carbon dioxide. The gas bubbles get trapped if you add flour, lifting the dough and making it porous. The more yeast cells, the more nutrients and flavor. But yeast cells stop growing when they run out of oxygen, or get choked in their own carbon dioxide. So I punch the bread down and stir it once in a while to let off excess carbon dioxide and put more oxygen in. Then I let it rise again before baking in a greased pan until brown on top. That's bread. If I want variety in flavors, I add things to the dough before baking—carrots, zucchini, apples, raisins—you name it.

My bread today was baked while I was writing this. I had mixed it up in less than five minutes. Today I had put in my leftover oatmeal, the remains of a can of condensed milk that I had not used up quickly enough, and a little whole wheat flour for variety. The product needs no filler such as cheese, meat, or lettuce to make it a meal. All it needs is butter.

October 11–13

What did I do the last two days? Perhaps typically, the little things. I cut down and into pieces the white birch with the fungal canker sores by the cabin, leaving the big healthy one. Since I had the chain saw already out, I cut up a couple more logs. I sat down to write, but it is hard to tell for how long. Hours can fly by as I write a few paragraphs, and I do not even notice. I went down to the phone to talk with Erica and Stuart, then came back up and pruned two apple trees. A hazy sun came out momentarily, and I took a walk through the maples and apple trees that I'd thinned out. I was awestruck in the maple grove, immersed there in the luminescent yellow all around. Yellow above me, yellow to the sides, and yellow below. It was a dazzling golden glow.

Some of the apple trees that I had freed from the embrace of red maples were bearing fruit and I took some along. They were perfectly formed, with no blemishes, no insect or fungus damage. You'd think they had been sprayed. Ordinarily, I like my apples with a worm in the center—it's the most honest advertisement there is for unsprayed fruit, and a little caterpillar doesn't affect the taste. The larva sits in the core, possibly making the fruit mature slightly faster. I've tasted a caterpillar or two on purpose. They're pretty bland, nothing to get excited about.

Last evening I climbed my observation tree to survey the fall panorama one last time. There was no wind and there was an overcast gray sky, but the air was crisp and clear and you could see the hills and the peaks of mountains fifty miles away.

A raven on liquid strokes came up from the direction of the silvery lake to the west, set its wings, and circled lazily down behind the dark green spruces where its companions were already feeding on the meat I had provided. I listened to the raven songs, and my eyes followed others in their erratic journeys all over the pastel landscape. Time stood still.

It could have been a half hour, or an hour and a half. I realized that I had watched and experienced as a child for the first time in a long, long while, maybe for the first time in my nearly five months here. I had not thought of the past or the future. I was mesmerized by the moment, and that liberation from past and future had enfolded me in the awesome beauty all around. Past and present telescoped to the immediate yet timeless sensation, the self evaporated, and I felt ever closer to the ravens.

Gradually I descended the spruce tree and slowly crept toward the feeding birds. The damp leaves muffled my footsteps, and I advanced slowly, always looking and listening, like a cat on the hunt, maybe for an hour, maybe for several. I saw their black silhouettes flashing through the trees and heard their singing pinions as they flew over me to and from the bait. Strangely, as I stood still and the birds flew closely by, they seemed unperturbed by my presence. Whenever a raven sees something of interest in the woods, it violently backpedals on rapidly beating wings, swerving right or left. But none of the birds coming near me did this. Had I become invisible to them?

Finally, through the screen of fir branches, I saw their black shapes dancing and jumping around the food pile, and I wondered if Jack was among them. I advanced a few more yards and at last they rose in a wild clatter of wings. But instead of flying off, they perched in the trees all around me.

FLYING SAMARAS AND CONSCIOUSNESS

The leaves have all fallen, but on windy days you see instead hundreds, maybe thousands, of little rotary blades spinning through the air. In large swarms these sugar maple seeds swirl along with the wind currents. The seeds are everywhere, and every gust releases another great crowd from every large sugar maple tree. The red and silver maples had already shed their seeds in late May and early June, before their leaves had fully developed. (Strangely, I would not see a single sugar maple seed the next fall, but the ground then had a carpet of their seedlings from this year's crop).

Seeing these impressive flights, I picked up a seed and examined it, noticing for the first time what a marvel of creation it really was. A maple seed is heavy in comparison to the seeds of some forest trees. The parent tree has packed in sufficient food to give the emerging offspring a start. This food offering allows the seedling to take the first, most critical step in its life. One big constraint, however, in this high parental investment is that the heavier the seeds are, the more they fall straight down. Directly in the shade of your parent is not a good place to try to grow up. Maple seeds are a study in compromise. They are equipped with a modest, but not extraordinary, supply of nutrients. They fly away from the overbearing shadow, but not very far. But nature has probably gone as far as it can in giving them wings.

The poplars and fireweed have gone all out for flight, compromises be damned. Their embryos are extraordinarily small so that the wind will carry them on their frail parasols. As a tradeoff, however, their seeds can grow only where there is no competition and where they can tap into the energy they need immediately upon sprouting. They can only grow in the extraordinarily rare circumstances where there is open soil in sunshine and where, at the same time, the emerging root can instantly tap into water. Otherwise they shrivel up and die almost instantly. They are the first colonizers after a burn.

The acorn, at the opposite end of the spectrum, can sprout almost anywhere, but it can't fly at all. It is dispersed by being carried away and buried by blue jays and squirrels. But for this service, it now runs an excellent chance of being eaten.

The aeronautical engineering principles that have been harnessed by maple trees throughout their evolution to make their relatively heavy seeds fly are not very obvious to me. All of the seed's weight is concentrated at one end, and the "wing," a thin blade, is at the other. You'd think the seed would fall straight down, like an arrow with a heavy point at one end and a feather vane at the other. What causes the seed to rotate like a propeller? I do not have the answer. I dropped seeds in still air, and they invariably fell straight down for a foot or so, but then got caught by the air and started to twirl. Once twirling, the wind held them and drove them along. Since wind is needed to carry away the seeds once they start twirling, the tree undoubtedly has a mechanism to release its seeds when the wind blows.

You'd think that such a fascinating phenomenon that is accessible to everyone would be researched to death. But checking with knowledgeable people at the Vermont Maple Laboratory, I found that the literature was practically nonexistent. I did learn that the seeds that were flying through the air are called "samaras," and that when two of them are joined, as they normally are before being shed from the tree, they are called the maple "fruit." According to the old literature, these samaras can fly a distance of at least "five chains" (198 feet). In one study in northern Michigan, it was determined that 70,000 of them fell per acre in the center of a ten-acre clearcut. But apparently nobody determined *how* they flew there.

In the spring I picked up silver maple samaras from the parking lot along the river behind the Farmington Diner. They twirled just fine in a

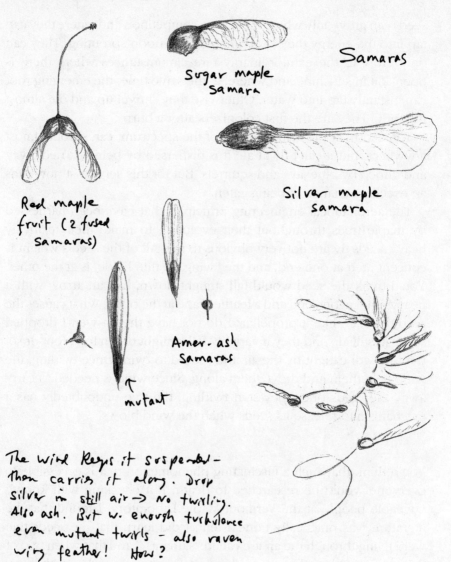

Samaras

Sugar maple
samara

Silver maple
samara

Red maple
fruit (2 fused
samaras)

Amer- ash
samaras

↑ mutant

The wind keeps it suspended—
then carries it along. Drop
silver in still air → No twirling.
Also ash. But w. enough turbulence
even mutant twirls - also raven
wing feather! How?

breeze, but dropped like an arrow if released carefully in still air. Red maple seeds, on the other hand, twirled even when released in still air. Thus, red maple seeds can fly in a very slight breeze, but silver maples need wind. My preliminary trials showed that, once twirling, however, the silver maple samaras may fly farther than the red.

Was the *rib* along the side of the samara important, since all maple samaras have one? I flight-tested ash samaras that lack the rib, but they, too, twirled in the wind. I found one mutant ash seed with three adjacent wings and even this one twirled, but only in a very strong wind. Aerodynamic engineers might know the answer, but if they concluded that bumblebees can't fly, what would they make of maple samaras?

The capacity to wonder allows us to anticipate, and that is a very big adaptive step. When you "see" something move physically before it occurs, you can prepare for it or even prevent it from occurring.

We all have the capacity to wonder. We all use it. But most of us must restrict it to the immediate things that affect our well-being. I am currently living a life of luxury. I can spend hours per day wondering about "useless" things, like the tri-partite feather vane on an arrow (rather than one blade for a wing), like how a samara twirls in the wind. The capacity that I have and that I ordinarily must use to try to find, make, or capture sustenance for myself and a family, has been temporarily lifted from my shoulders.

Evolution never produces anything "extra" (except incidental by-products), because everything it produces has a cost. Our extraordinary capacity to wonder is a reminder of how much survival and reproduction had at one time been dependent on the ability to visualize. If, as the studies of many so-called primitive societies suggest, the making of a living close to nature was not difficult, then the ability to visualize was probably related to anticipating the moves of others, and possibly related to sexual selection. A robot can outcompute a chess master—big deal. But can it court? Could it bed and wed?

There are some who argue that consciousness, even pain, are "extras." They say that you can make a robot that recoils from a wall or hot stove, yet feels neither pain nor any conscious awareness of its acts. Why, they then ask, do *we*—specifically—have consciousness? Why do we feel pain, hunger, thirst, love, hate, jealousy? The answer is not, I believe, as difficult as some philosophers would make us believe. We have natural experiments to show us. There is a rare medical condition in which a person is incapable of feeling pain. Such a person is not afraid of battering himself or herself until bruised and cut. In short, pain is a warning system that, coupled with learning from experience, makes us avoid that which might damage us. Consciousness is adaptive to such complex creatures as we are, because if we feel the pain of a pinprick, it spares us the trouble of driving a spike into ourselves to find out whether or not this hurts us also. Similarly, after we jump off the back steps, we no longer need to jump off the roof, to find out if that hurts. "Unconscious" birds and mammals may appear to plan ahead, but act according to hard-wired responses. All might work fine, until one actually *did* plan. Then the others would be disadvantaged. American Airlines can no longer afford *not* to have DC-10s and 747s, even though the Twin Otter or the Piper Cub can carry passengers, if the other airlines have the newer planes. It is not what works that *alone* matters; it is what the competition does that is crucial. Consciousness allows us to visualize the consequences *before* we have physically experienced the cause, and a person without consciousness would be disadvantaged relative to one with it.

Does an angleworm feel pain as it writhes upon the fisherman's hook? Very doubtful. A worm is not subject to mechanical injury in its normal life, and one with it gains no advantage relative to one without. A worm must, however, be able to avoid obstructions, hence it writhes in the same way as our knee reflex works. Does a lobster feel pain when we throw it in a pot of boiling water? Again, very unlikely. There is no reason to suppose that evolution would equip it to avoid a kind of danger—temperature shock—that the species has never experienced in the last 100 million or so years of its evolutionary history. (A whale would likely feel this pain. Its ancestors were land animals that had to be able to deal with temperature extremes, and it still regulates its body temperature.) A lobster likely does not feel the heat, either consciously or unconsciously. The worm feels the hook, but unconsciously, hence it is not "pain."

Some mechanisms are inordinately simpler than others. A male moth flies upwind to a scent, and it goes through a very complicated repertoire to do it. But it need not be conscious to do so, nor does it need to feel pain or pleasure. Why do I make this bold guess? Because a moth's decisions are binary, like those of a computer. The moth's antennal receptors pick up a scent molecule. That is the alert signal, or "on" switch. The question is now whether to go: "yes" or "no." A "no" might apply when it is too cold, or perhaps too early in the day. A "yes" means "start to shiver." Shivering stops at a temperature set-point that automatically activates flight. Flight is upwind if scent is detected, zigzagging if not, until scent is detected again, and so on until the female moth is reached. A similar set of responses that rarely vary is instigated. Whether the moth feels anything or not makes no conceivable difference in the outcome. Therefore evolution would not load it up with such extra baggage.

So why is it *not* extra baggage in a human or in a dog? Most of our decisions are *not* binary, unless we are playing chess where a specific goal is desired. When I "merely" step out of the cabin, I may or may not have a specific goal in mind. Any one decision to step out the door, however, is one I have never ever before made in my life, and it is a decision I will never ever have to make again. Before going out, I have to decide among hundreds of alternatives. Should I stay and sweep the floor? Wash the dishes? Write? Answer letters (which ones)? Do a drawing? Read (a hundred options)? Should I go running now? Check on my answering machine? Look for a wren's nest? Cut brush? Make firewood . . . ? Every footstep I take consists of a compromise of untold hundreds of alternatives, each one with innumerable consequences that I should be aware of. To make the optimal decision, I visually scan as many as possible of the various alternatives. This scan may only take a few seconds, but it is made nevertheless. During this scan, I visualize myself in each or many of the situations, and I weigh the likely result of each alternative. The weight of the alternatives shifts constantly—as more dishes pile up, as I survey the field and see chokecherry leaves that I envision turning to trees in 15 years, as I hear a wren's call, and so on. No binary decisions are possible, because there is no endpoint to my actions. I am free. You can program a computer to do *something*. Take that "something" away, and it has no more decisions to make. It is dead.

If, after all this, I decide that my step out of the cabin will be in the direction of a rendezvous with Jane, then in order to have any glimmer of success I must consider what Jane is doing at the moment, how she has been responding lately, what has been of concern to her, where she might be now and why. I cannot begin to fathom how she will respond to "Let's go eat a duck," "Let's climb Mount Katahdin," "How about a date?" or "How is your mother?" unless I can imagine myself in her shoes, so I know how she might react, before she actually does. The more I can visualize or imagine myself in her shoes, the more accurately I can predict and respond appropriately. In short, consciousness or visualization is not extra baggage at all. For us, it is a *necessary* ingredient, perhaps in no small measure because we are social animals. Is my lack of social interaction now diverting my consciousness from its primary evolutionary task of predicting and hence existing in the social environment?

And love? It consists, according to Saint-Exupéry, "not in gazing at each other, but in looking together in the same direction"—in practical terms, in having a common goal. Is love necessary? To the moth, it is not: its tryst is so brief, and its life so short, that pure sexual attraction is sufficient to achieve all that is necessary. Anything more is totally superfluous and it would therefore not have evolved. Indeed, even the sexual attraction itself likely is not conscious in the moth.

Of the many kinds of love, however, all relate to *social* interactions: between parent and offspring, offspring and parent, partners, comrades, or mates. In all cases, love is the glue that binds the relationship together. And the only reason why evolution would bind relationships together is if they served a utilitarian purpose. Otherwise love would be superfluous, and would not have evolved. The moths have no need to stay together. But the parents of many birds and some mammals need to take care of the young for a long time to ensure that they'll survive and that the parental investment is realized—therefore, bonding is necessary. The offspring obviously need their parents, so they have evolved to love them in order to maintain the closeness required for support. Any partnership requires love, whether it is for child rearing, for economic reasons, or for any other mutual project. The more independence the "partners" have, the less their need for love, and the less you'd expect it to flourish from the evolutionary perspective. The more I come to depend on the forest, the more I'll love it. I have depended on the forest now for many things for a long time, even for the entertainment of watching samaras in the wind.

October 14
TO FIND SMALL CATERPILLARS

When Jack first left the nest, he spent days at a time picking up and tearing apart everything he found. He picked up and shredded every other leaf. He ripped up grass; tore apart moss; picked up pebbles, sticks, and twigs. Gradually he became more selective. He tore out the chinking in the cabin. He went for contrasting objects of regular shape, like cups, spoons, pencils, catkins, spruce cones, and flower petals. And then he discovered little beetles. He ate them and soon dropped other toys. He learned to distinguish the new and the relevant from the background milieu of his environment; at a glance he could pick out a beetle, a fly, or a caterpillar worth eating from the incredibly rich and complex background of a meadow or a forest floor.

I was starting to do the same thing. After long and daily immersion in my surroundings, I noticed with shock that I barely stopped to listen to the hermit thrush. The bird's beautiful song became background, to be mentally shed in order to hear the new. If there was a tiny squeak to one side of the trail that I had not heard before, then my head snapped involuntarily around in one instant and swift motion, even before I was conscious of the sound. I had responded *before* I was conscious. My mind still noticed all the common things but no longer focused them into articulate thoughts. I had become like the raven.

BURNING WOOD

The hills are gray now, because the hardwood twigs and trunks are bare. Here and there is a tinge of russet brown, and from a distance the forest is finely etched in thin vertical white stripes—the white birch trunks show up like candles in the dark.

The wave of defoliation came from the north. It passed, and is now rapidly advancing south. The color of the fallen leaves has drained just as quickly. After only a few days, there was not a yellow or a red leaf left among them. They are now all in different tans, browns, and reddish browns. Soon they will all be dark brown.

Cutting this winter's firewood means that I've thinned out my subsistence woodlot to make some beautiful maple, birch, or ash grow faster into big trees. It also means I've got warm comfort in the bank, and I've had a good workout. Finally, it also is tangible satisfaction when I get around to using it because I remember the work put into it.

My small upstairs stove throws out a lot of heat and is probably fuel efficient. I can close it down at night and often still have hot coals in the morning, but my ancient cast-iron kitchen stove downstairs, complete with an oven and a large flat top, is wonderfully inefficient. I say "wonderfully" because it throws off little heat. I can heat the oven and the cooking surface just above the flame, but most of the heat goes up the chimney. This is just as I want it, in the summer. The stove is not airtight, hence all dry wood I put in is burned completely, "efficiently," and with little emission, within an hour.

Efficiency can be measured in more than one way. My kitchen stove efficiently *burns* wood, but it very inefficiently heats the cabin. If I close down the flue up the chimney, then less of the heat escapes up and out. Totally closed down the stove would be close to 100 percent efficient in transferring the energy from the wood to heat into the cabin. But the house would then be full of not only all the heat burned from the wood, but also all of the smoke. I burn the wood more efficiently (completely to carbon dioxide and water) if I give it more access to oxygen by opening the bottom vent of the stove, but while that creates less smoke overall, it would dump more smoke into the house if I also closed the top damper to the chimney.

Like baking bread, you have to know what you are doing to operate a stove. But I don't have to think about it consciously any more than I have to think in order to accomplish the inordinately complex task (involving binary decisions) of digesting a peanut-butter sandwich. I did not realize there was anything to burning wood, until I had a friend from the city stay here one day alone. When I came back to rescue him, he was coughing and shivering. The whole cabin was full of choking blue smoke and there was still no fire in the stove.

SCENTS AND SOUNDS

Last night a crowd of ravens slept again in the pines not more than 100 yards from the cabin, as they have for many nights now. At these nightly gatherings, the birds sometimes burst forth into the wildest, most incredible jubilation, that lifts even the most jaded spirit. This time they were quiet except for an occasional short outburst of bickering. I can read many of their moods, not so much by the specific calls they make but by their tone, in much the same way as if I were at a bar and the fellow drinking next to me, talking only in Swedish, had just won the Megabucks lottery ticket. The raven celebrations are probably similar; maybe the birds have been hungry and, when one of their members discovers a moose carcass, it vents its feelings. The others then know it has won a lottery and, knowing they'll get some of it tomorrow, they get excited, too.

Deep in the woods there are mixed flocks of chickadees and kinglets, and they chirp incessantly though unobtrusively. Unless I am alert and purposefully listening *for* them, they do not enter my consciousness.

The sounds of the weather are like that, too. There is the gentle pattering rain on damp leaves. There is the faint rustling of dry leaves on sunny days, and the hum of the conifers as the wind rakes through them.

Almost every evening, shortly after dusk, a barred owl lets loose with an ear-shattering and eerie series of calls that would frighten me if I did not know what it was, and that I expect strikes terror into the sleeping ravens. A few nights ago, a coyote began a slow, low howl that increased in volume and pitch, switched to a high tremolo, and resounded back and forth in a harmony that left me in awe. Usually these solos by individuals are lead-ins for a whole orchestra of others who fall in with their own compositions, all perfectly harmonized. As the volume and the excitement mounts, there soon is yipping and barking, and you'd think you're surrounded by a pack of 50 of these wolf-dogs. Then, as you listen closely, you hear an answering pack from a distant ridge. And then, abruptly, all is silent again. Sometimes one howls alone and none answers.

The smells rise, too, in late fall. There is a sharp tang mixed with an aroma reminiscent of fresh walnuts and hazelnuts. The smell is faint and subtle, but it powerfully affects my mood. It is like a channel to my childhood memories, and I strain to inhale deeply whenever I sense that earthy aroma of fallen leaves. The memories it evokes are often not specific, but sometimes the bittersweet flashbacks are so sharp they almost hurt.

Yesterday evening as it was getting dark, I walked the path past a large sugar maple tree that had shed its leaves. It was drizzling very lightly, and I could hear the tiny patter of small raindrops. But as I passed the maple, I smelled *that* peculiar smell of autumn—just a quick whiff. I stopped to catch more scent and my mind flashed back to 43 years ago as I saw myself following my father on a dark evening. It was cool and damp then, and the leaves undoubtedly had given off that same aroma.

Papa was tending his mouse traps, to catch wild rodents and shrews to be sold to museums in America. These excursions in the dark were like a door to another world—the secret world of mice and shrews and moles and voles. He set the traps carefully under mossy logs, under grass overhanging like curtains along steep banks, and in brush piles. In each different place, he caught different furry creatures that I would never have known existed. They were creatures that could not be seen any other way, for they led hidden lives under the cover of moss, matted grass, and leaf mold. At that time the excitement of our explorations into this unseen world had set my mind aflame with pleasure. And now the magic wand of scent had lightly touched, and passed.

At first it seemed that something had been lost, never to return. The thought touched me with a tinge of sadness, at the same time that the scent touched me with happiness. Yet, thoughts create feelings, and I then had another thought that made me feel better.

The maple tree beside me is 99 percent dead. Its only living tissue is a thin layer of cambium just underneath the bark. Each year the cells of the cambium divide, and those that align themselves toward the inside of the tree die and become wood, its support. The autumn scent had given me access to the deepest growth rings of my life, which served a vital function—support for new experiences, and new growth.

TREES

It rains off and on today and stays overcast with only a very slight breeze. I split and pile wood, and pick the apples that have fallen off the tree near the cabin. They are all deformed, but I cut the good portions out of some to make an apple crisp. (The rest were used earlier in the apple toss contest at the annual rendezvous. In this contest, you stick an apple onto the point of a long, supple pole, and then you whip the apple off to new heights or distances.)

While I'm at it, I also bake some bread, which energizes me to work two more hours swinging an ax, thinning out another abandoned field that is overgrowing into a thicket of ash saplings. I'll be long gone by the time those trees grow up, but in the meantime I'd like to watch them grow. So I do something for the "right now," something for weeks or years from now, and something for the future.

At this moment, I'm having a cup of coffee, listening to the rain pattering on the roof, and smelling freshly baking bread. My cabin is made from logs of fir and spruce. My table is made from pine boards with cherry legs. The warmth that I enjoy inside on this blustery day comes from the burning of red maple and ash. My eye feasts on the panorama of forest before me, and my mind gathers tranquility from the thought of the birds, the squirrels, the mice, the millions of insects, and the other animals that live here. The deer, the moose, and the snowshoe hare browse on the trees' twigs, and the ruffed grouse feed on the buds, and I feed on these creatures as necessary. No wonder some cultures worship trees. Whether we are conscious of it or not, we are children of the forest. Thoreau heard the "souing" of the pines, as if they were animals moaning in the wind. And he stood upon a large pine stump in Maine and railed against greed and all the evils of mankind, as if by cutting down that one beautiful tree we had cut all trees. Thoreau saw beyond most of his contemporaries, but maybe we need to see further still.

There are some things that we need and some things that are merely useful. Computers are useful for writing, but I don't need one. My pen works fine. I can write fast enough, and there is no rush. Nuclear power plants can be useful, too, but we don't need them in Maine, where the population is not yet bursting beyond what the environment can support. Windmills might be preferable to nuclear power plants, but trees are preferable even to windmills. And we need trees.

Engineers have recently experimented with the concept of capturing solar energy. But trees have been *doing* it for hundreds of millions of years—billions, if you include the construction of the miniaturized solar batteries called chloroplasts. They've assembled themselves into the ultimate: giant, multilegged structures, some up to 100 or more feet tall. They are nonpolluting, and indeed reduce pollution rather than create it. They assemble themselves, repair themselves, and require no maintenance.

Trees as an energy source also have the advantage that these energy "factories" can be installed almost anywhere. They can be grown intertwined in cities. They can be installed on steep hillsides. They can be put onto floodplains, and even in deserts. But, above all, they can *make* forests and wilderness, at the same time that they are making energy. Purists object that once you have a forest, it should be left inviolate. We need to act logically, but we also need to act bio-logically. By refusing to make intelligent use of our forest resources, we insure the need for hydroelectric dams, oil pipelines, and more intrusive installations.

As a card-carrying member of the Wilderness Society, I do not advocate less pristine forest. To the contrary, my plea is for more. But if it comes to a choice, I choose *forests* as a power source over dams that flood forests.

October 26–30

Winter is coming. I heard it last night. A moaning north wind that ebbed and flowed like the sound of surf and ocean waves. Now, in the dawn light under gently drifting clouds from the north, the ledges on top of Mount Bald are suddenly clothed in a sheet of white, and the dark spruces below the peak are framed in white. This morning as I get up, my watch says 5:45 AM at dawn, rather than 6:45 as it usually does. I had heard that we were supposed set our watches back one hour last night, so I did, just to go along. "Fall back, spring forward . . ." Whatever. Snowflakes swirl around in front of my window, and slowly, imperceptibly, they mat down the leaves and the grass. The cabin feels chilly, and as I start up the woodstove I tell myself to haul in even more logs from the forest. Winter is coming on the north wind, and winter in the Maine woods comes to stay for six months.

This is a good time to walk in the woods because with the leaves down, it seems like a curtain has been opened. You can see for a long ways, and where you would otherwise have been enclosed under a green ceiling, you can now look up and see the hills and mountains. The tall oaks in the hardwood forest still have their tan-brown leaves, however, and although the mature beeches are bare, all the young beeches below the forest canopy still hold their now dry pale tan leaves. They will keep their leaves all winter. So will the sugar maple seedlings, for a while, forming a yellow carpet in the hardwood forest.

The fallen leaves are already curled and drying, and they now let you hear every creature that walks about. A small quick rustle, and I get a quick glimpse of a mouse's back. The mouse does not reappear, but in several directions I hear more steady rustling. Red and gray squirrels are foraging. Everywhere there are chipmunks with bulging cheek pouches, gathering sugar maple seeds that have recently fallen. The striped maple seeds are now also beginning to fall. But the ash seeds are still up, as are the birch seeds; both will be vital food for the finches this winter.

In the mixed hardwood forest, I come across a number of hairy and downy woodpeckers. But suddenly off the ridge comes a much more imposing, big black bird, with white under the wings: a pileated woodpecker. It folds its wings, and down the hillside it shoots past me, opens its wings to catch the air with a zooming sound, then folds them again and shoots further down the slope. Now it extends its wings fully, catches the air, and in a quick upward swoop lands in the top branches of a basswood. There it cackles a loud, penetrating call that reverberates off the hill and across the valley. With jerky motions the cock of the woods, as it is called, looks this way and that, and remains rooted to the slender branch for five minutes. Then it launches itself and swoops down the hill even further. I do not see where it lands, but soon I hear a low, loud drumming. The drumroll lasts a second or two. Strange, I think, because woodpecker drumming is supposed to be a spring activity, associated with mating and sexual signaling. Another drumroll. This one is of a slightly higher pitch; the bird must be drumming at a different spot on his chosen tree or branch. And then another drumroll—that's three. Now I'll start counting.

I did count, and the bird made 18 drumrolls, each from two to five minutes apart. By the 18th roll, I was already down the valley, and it is possible that the drumming continued. I have no explanation for it, and it reminds me that no matter how much I know about birds, there is always more mystery. And despite all my scientific explanations, I know that, fundamentally, they sing, drum, or dance out of simple exuberance. This woodpecker, in every action that I saw from its flight behavior to its idling in the crown of a basswood tree, was overflowing with health, exuberant, and its spirit was contagious. What more would it signal to a rival or a mate? I, too, felt light in my steps and free in spirit as I continued down the south-facing hardwood slopes of Houghton Ledges.

I came to the site of an old farm that had been overgrown and then clearcut five years ago. Scattered through a fast-growing thicket of ash, oak, sugar maple, beech, and striped maple, along with patches of raspberries and blackberries, were apple trees that, now unshaded, had taken on new life. They were loaded with fruit. I was surprised to see that *all* of these apples, just like all of those that I had released from the forest near my cabin, were perfectly formed and unblemished with fungus scabs or insect damage. They likely had not seen any spray for their entire existence of perhaps 100 or 150 years.

I was not the only one enjoying the apples, nor the new dense growth of trees. The thick slash lying everywhere had been packed down, and decay had made it brittle. Trails had been broken through the brush by large animals. There were piles of apples on the ground, but one of the apple trees had large limbs freshly broken; a bear had been there harvesting fruit even before it fell naturally. The ground was freshly pawed by deer, and the smooth black earth showed the criss-cross pattern of their hoof prints. Piles of their black pellets were strewn about, and here and there were fresh mounds of bear dung that looked like applesauce.

In hushed silence I stopped and listened. A grouse clucked softly in a thicket nearby, then I saw it run ahead of me. But the deer and bear were not to be heard nor seen. They probably come here only at night, traveling quickly through the open, still uncut forest that I myself had just crossed. Possibly they had gone down one of the trails through the brush after feeding here, to spend the day on beds of sphagnum moss down in the balsam fir and spruce swamp. It is close to hunting season, and they may know when their chief predator stalks.

Descending further through the clearcut on one of these trails strewn with more deer and bear dung, I went past blackberry vines that undoubtedly had fed the bears before the apples ripened. Everywhere the tops of the shoots of the young trees were browsed off. The more they were browsed off, the more side shoots would grow next year. Thus, for a while, increased harvesting would generate more food. Meanwhile, some of the shoots were escaping the browsers: first the snowshoe hares, then the deer, and lastly the moose. These escapees would grow all the faster because their competitors had been pruned. No crew of men was needed for forest improvement—the animals were doing it themselves.

Walking through a thicket along a game path I suddenly saw a cat-sized animal, jet black with short legs and bushy tail, dart off in front of me. It was a fisher, an animal related to weasels, whose tracks I've seen often in the wintertime. It had come from a heap of old slash that had been piled on a boulder, creating a cave along its side. And there, just in front of this cave, lifeless and on its back, lay a young porcupine. It did not have a trace of rigor mortis, and was as warm as my own body. Fresh blood oozed from puncture marks at its throat, and nothing had been eaten yet. Only a tiny bit of skin had been torn off the inside of its right front shoulder. The fisher is the only predator known to be able to subdue the lethally armed, spiny porcupine, and I had always heard that it does so by attacking its victim's soft underbelly. But this porcupine had no scratch on its belly; it had been attacked at the front end instead.

The porcupine's lethal weapon is its large, muscular, spine-studded tail, with which it lashes at its antagonists. There were signs of a struggle: inside the fisher's cave, one log was impregnated with a patch of sharp white quills, which are barbed like miniature fishhooks. A powerful tail-swipe had missed its intended mark.

Leaving the battle scene, I was soon in the spruce-fir swamp. The ground was covered with soft green sphagnum moss, and the dampness here did not permit the growth of hardwoods. A troupe of golden-crowned kinglets was foraging close to the ground. I stood still and two of them came within five feet without appearing to notice me.

It got increasingly cloudy as I started walking back to the cabin, and by evening, when I returned, it was drizzling. I built a roaring fire in the stove and baked some potatoes.

In the night I awoke to the honking cries of Canada geese. I jumped out of bed and stood in front of the cabin. They were coming from the north, flying directly over the cabin. Their wild, excited calling sent shivers down my spine as they continued on their journey south.

ELECTION TIME

The sun warmed the gray weathered logs of the cabin. A few cluster flies found the warm logs and sat on them. Others joined them, and soon there were hundreds. Then, as the air got cooler, they crawled indoors. The warmth inside made them lively, and they started bouncing against the closed windows. Their combined din, after a few hours, got to be annoying. I gathered them up with a loud sucking noise, using the battery-powered Dustbuster, then etherized and censused them. The last step might seem superfluous, but it occurred to me that you might ask, "Just exactly what does he mean?" when I report to you that I had "a lot" of flies in the house.

I spread the etherized lot out on a newspaper, and then I carefully counted 100 out into one pile and divided the rest of them into roughly similar piles. It came to a total of 31 piles, or about 3,100 flies, which is roughly "a lot." But as I'd learn later, this lot was just a smidgeon.

These cluster flies, genus *Pollenia*, get their name because in the winter they aggregate in big groups in the warm protected crevices of houses. Whenever the house is heated, they come out of their hibernation clusters and fly toward the light (windows) to try to escape, as if spring has sprung, to search for the haunts of earthworms. The larvae are parasites on earthworms, and each adult means one dead earthworm. Our common earthworms were introduced from Europe, and so were at least two species of *Pollenia*. The imported *Pollenia* flies are calliphorid or "flesh flies," easily distinguished from other local calliphorid flies, which are shiny green or gunbarrel blue. The *Pollenia* flies are dull black, like houseflies, but on the thorax they have crinkled whitish hairs among longer, erect black ones.

Charlie, my nephew, has come from North Carolina. He's taken a week off from his graduate studies to go deer hunting with me, a somewhat more traditional activity this time of year than counting flies.

On opening day we hunted within walking distance of the cabin. In the evening, we sat at the edge of the clearcut, hoping to surprise deer coming down off the ridges to feed in the open. The leaves were dry, and the woods were noisy. Now and then you'd hear a quick rustle, and just barely get a peek at a red-backed vole breaking the surface of the leaves along a log. I became attuned to the high-toned squeaking of shrews, which appeared to be engaged in conflict. But I saw none, for they seldom came up above the leaves. Chipmunks scurried over the dry leaves in short bursts, sounding like moose on the run. Here and there a red squirrel jumped down noisily, then hopped back onto a horizontal log cushioned with soft green moss. Towards twilight, the gray squirrels made their rasping coughs, and you'd see a gray shape dart up a red oak trunk in the distance. The leaves amplified the activity of all the creatures, including the birds, as blue jays landed occasionally to cache beechnuts and acorns. One hop of a jay or a chipmunk would be as loud as a footfall of a deer, and so our senses were on high alert to try to figure out the patterns of sound from far and near. Near dusk I finally heard one steady, slow, rustling noise approaching, which sounded different from all the rest. I saw its dark, lumbering shape, low-slung and squat, coming from the clearcut—a porcupine.

On the next day we explored an overgrown apple orchard on the Byron road, finding many fresh buck scrapes and pawed ground. Here, we each picked a good pine tree to climb, where we'd come back to sit the next morning. We got up at 4 AM, and by 5 AM (still in the dark), we were already in our trees.

In the first gray dawn, I heard the alarm calls of robins. Two landed in an apple tree beside me on the pine. It must have been cold, because they were both fluffed out into balls, with only their bills sticking out from one end and their tails from the other. Their legs and feet were completely enclosed in their breast feathers. Below me I heard a rustling. At first it was too dark to make out the small animal, but eventually I made out two white-throated sparrows. They were foraging among the fallen apples and apple leaves, by jumping forward in little hops, each time kicking back, thereby disturbing the leaves and exposing the ground underneath. The two robins soon left their perches beside me and joined the sparrows. Instead of kicking the leaves, however, the robins flung them aside with their bills in quick jerking motions.

A small flock of evening grosbeaks flew over, and at first light I heard chickadees and goldfinches. Later, long after it was light, a surprise came by: a palm warbler. This warbler breeds in black spruce bogs. This was not its habitat, and it was obviously coming through on migration. I was surprised to see a warbler so late. My eyes feasted long on its yellowish, olive green–streaked plumage. Curiously, it wagged its reddish brown tail up and down rhythmically, as does at least one other quite unrelated bird, the wagtail, and another one, the phoebe.

A red squirrel was another early riser. It scurried along a horizontal log, stopping to nip off a small fungus, sit on its haunches with the fungus in its front paws, take off a few bites, drop it half-finished, and continue on, doing the same to a second fungus, and a third. Then it dashed to the ground near me, dug under a leaf, and pulled up a damp spruce cone. Back onto the log it hopped, to hold the cone like a corncob and chew round and round from the large end, snipping off the bracts to get the seed under each one. For a few seconds the squirrel was behind a few pine twigs that were blocking its view towards me. Back and forth it dashed nervously, until it again had me in full sight, and then it resumed its meal with its left eye toward me. At the side of its head, a squirrel's eyes are suitably placed to scan in all directions, but they could not possibly see to direct the delicate maneuvers of chewing off the bracts and picking out the seeds. These things had to be done by feel alone, or by dead reckoning. All the while, the squirrel was less than ten feet from me, and although it kept a wary eye on me (or perhaps a curious one), it did not seem alarmed.

When gray squirrels live in cities, you can feed them out of your hand. But here in these woods, gray squirrels, in contrast to red, are one of the shyest of animals. You can see them on the hardwood ridges, but only if you sit down and wait patiently. After a while, maybe ten or twenty minutes, one of them will begin to scold. However, if you take one step, it will hide behind the trunk of a tree instantly. The squirrel will track you by ear, always staying hidden by pressing itself flat against the tree trunk opposite you; even if you walk around the tree that you saw it run up, you'll not see it. I have seen a buck whitetail deer act similarly by hiding behind a boulder and crawling on its belly around the other side as I walked by. The only thing that gave it away was a sapling curiously moving on a day when there was no wind. The buck acted as though it "knew" it could be and had been seen. Does this reaction suggest a heightened consciousness, beyond mere reflexive responses? Can the deer or the squirrel visualize itself from the point of view of its pursuer? Is it this ability that allows these *shyest* of creatures, along with the raven and the Canada goose, to become tame and to adapt so easily to contact with our civilization?

We descended to the ground after about two hours, and then we walked for a while up and down and around the ridges of Potter Hill, near the north end of the lake. Near the upper west-facing slopes, I was surprised to find woods that seemed ancient and undisturbed. The trees were not large—none were more than two feet thick or more than sixty feet tall—but there was no evidence of rotting stumps where bigger trees had previously been taken. Many of the old standing trees showed signs of age and decay. Other large trees, of similar size to those standing, had fallen and were now slowly decaying back into the ground. They were covered with green moss. If there had been timber harvesting here, these now prostrate rotting trees would surely have been taken 40 or 50 years ago. Here was a natural progression of young trees, old trees, and decaying fallen trees. It was a beautiful mixed forest of oak, white birch, sugar maple, and spruce, with a sprinkling of balsam fir, beech, hornbeam, and American ash.

Further down the slopes of Potter Hill, there was evidence of lumbering. On one south-facing slope, we walked through a beautiful beech-oak forest. The trees here were all larger and growing much more vigorously than in the virgin forest above. But none were half-decayed or half-dead, as many trees should be in an untouched forest, nor were there large dead trees on the ground.

Still further down the slopes, we came to stone walls and ancient apple trees among a newly emerging forest. It was here that the ground was tracked up, as in a barnyard. And it was here that we spent the last three hours of daylight in the top of an old apple tree. This apple orchard was only now being reclaimed by the woods, hence it still yielded apples. There were apples of many varieties that I had never seen. Some of them were delicious. As on my own hill, none of these wild apples were worm-eaten; most were perfectly formed and not disfigured by fungus or borers. I made a note to myself to come back in early spring to get scions for grafting.

Throughout the orchard bordered by stone walls, the forest was making a dramatic comeback. It does not respect stone walls. Ash and sugar maple trees were shooting up among the apple trees in the remorseless struggle for light. It was a fight to the death which the apple trees would lose, having grown up pampered by the orchardist, who had kept the field clear of any competition. The apple trees had taken advantage of this easy life, branching their trunks in all directions, to try to grab all the light they could, over as wide an area as they could.

Eventually the orchardist left, and now they were on their own. By this time their crowns nearly touched, but none had felt the relentless competition for light, and it would be years before they would feel it. By the time they did, it would be years too late, and their fate would be sealed.

Since the orchardist had left, thousands—millions—of tree seeds had blown in from the forest. Sugar maple seeds, like little helicopters, had whirled in the wind and settled under the trees in the fall. For many years, other crops of ash and maple seeds landed and sprouted. The young seedlings had only one way to go—straight up. They were crowded acutely on all sides, and they died by the millions, every year. Thin beanpoles were left, groping upward through the lattice of the spreading apple tree crowns. A few of these very thin ash and maple poles finally reached the light above the apple trees, and now the apple trees were starting to respond. Each tree now sent its own shoots straight up, but each tree's crown was wide, hence it dissipated its energies into many shoots, whereas the beanpole ash and maples, disciplined early by the competition, had put all of their energies into single upward-striving trunks. Now, having reached the sunlight, they were starting to expand their own crowns, grabbing the sunlight away from their former nemeses beneath them. Now their trunks would thicken ever more as their crowns grew large. The apple trees that had grown for a hundred years would be dead in ten. I noticed that some of the upstart ash trees were already bearing seeds, and some were loaded with them. Pine and evening grosbeaks would feast here this winter. Meanwhile, a grouse landed in an apple tree beside me, then flew down to the ground and walked away.

We saw no deer all week. The deer are now nocturnal. We followed their tracks down into the swamp where a recent clearcut had left impenetrable thickets of young fir. Undoubtedly, they were bedded down here for the day. But to see one in this dense growth, you'd have to stalk within 15 feet of it. Fat chance.

On Tuesday, November 3, Charlie and I left for the woods before dawn, as usual. We spent most of the day together, walking slowly, stopping occasionally for an hour or so, then finally splitting up near 3 PM to sit the rest of the day in a tree.

Sitting on a limb in a slight drizzle, I felt detached from the present time. All my senses were directed outward, so my imagination was free to wander back and forth in time at will. I saw what was once a farm and fields before me. A cellar hole remains as a reminder. There were kids traveling to and from school down the country road through the woods. There were sleighs and carriages, there was apple-picking and hot cider on cold winter nights. I saw a little girl who grew up here in the woods—whose headstone I read in the graveyard below from her death at age 21. One moment this is a wilderness with wolves. Then a little girl picks apples here in the fall. Now I am sitting in one of those apple trees surrounded by a forest growing up around me. The curse of consciousness is to see death, but it also sees the process of the whole. And the process makes each death less lonely, because life always continues into something else. It never dies.

This was Election Day in the outside world. Charlie and I did not discuss it. (I had already voted by absentee ballot.) Possibly hundreds of thousands of people were glued to the edge of their chairs watching CNN for the latest update of returns from New Hampshire, New Mexico . . . But I could not see the point in spending hours trying to find out what would happen when it would happen regardless, and when we could get the results soon enough.

As we were coming home, we stopped at the village store. "Charlie, how about grabbing a can of red kidney beans for supper?"

"OK." Charlie jumped out of the truck and came back two minutes later with the beans, telling me of the election of the new president.

"Yeah," I said, "but did you remember to get some peanut-butter cups for dessert?"

November 8–9
JACK?

There is a thin crust from the snow that fell the night before last. Walking east, down towards the brook where the old apple trees are,

I saw deer tracks that were made the previous night. Lots of them. Heading north along the brook where the woods are dark from the thick firs and spruces, I heard a crash and saw a brief flash of white. I had jumped one. Of course, the deer had probably heard me coming while I was still a mile away.

I swung up around Chandler Hill, then along Gammon Ridge that is covered in hardwoods. No sign of deer here. But in a clearing created by logging, I came upon an assembly of ravens that were loitering about, pulling at sticks in the ground, pecking apart rotten stumps, and engaged in other senseless doodling. This was probably the group that had been feeding on the calves I'd provided near the cabin, about a mile away . . . as the raven flies.

At Houghton Ledges I turned south to come back home, and coming back up my hill I crossed the track of a yearling moose and also that of a bear. I was surprised to see the bear track this late in the year.

It was nearly dark. The ravens had returned to their roost in the pines and were noisily getting settled when I got back. They squabbled and occasionally trilled. I had been providing food to keep this aggregation here. Had they found something else? Clouds drifted over to veil the almost full moon, and I heard somewhere from Gammon Ridge a deep, howling wail. A wonderful baritone voice, full and rich, it gathered up the mysterious night into one theme. The slow, languid monologue was followed by yips, then by barks, and more howls. Other voices chimed in, some from my own hill. I wondered what the deer thought of these sounds in the night. I once saw one of these dogs peering at me through the fir trees. It sure looked like a gray wolf. About nine months later, a wolf—a black 67-pound female—was shot further north in Maine.

I went to sleep to another wolf-coyote serenade, and this morning at dawn I awoke to a slow, throaty wail that ascended and then descended in pitch, followed by rapid staccato yips.

At 3 PM the next afternoon, I climb my spruce tree to watch the ravens return to their roost. Visibility is excellent and I can even see Mount Washington, a 90-minute drive to the west. For an hour I feel suspended. Not a raven in sight. Near 4 PM the sun is low over the orange horizon of the White Mountains, and the white globe of a full moon is rising in a dark blue sky to the east. Black shapes now appear against the limpid sky on the horizons! There are singles, pairs, and scraggly groups of up to a dozen coming from various directions. In the vast bowl of sky bounded by mountains on all sides, the birds seem magical as they approach. Their slowly rowing wingstrokes achieve distance that seems out of proportion to the apparent effort. The sky—calm, cold, and clear—is their element, and it is hard for me to imagine a performance of greater elegance than their flight.

They come in at treetop level, and they also come in at great height. Those arriving high in the sky descend by banking sharply on outstretched wings, then spiral down. In their downward plunge, however, some pull in one wing to make a quick partial roll. One shoots down on two tucked-in wings, then in the flash of an eyelash does a full roll, then keeps right on diving. Although most fly languidly, several that pass by close to me on my lofty perch exert vigorous, rapid wingbeats that slice through the air like powerful propellers. As they fly by, I can see their heads alertly swiveling this way and that in rapid jerks, to the sides, up and down, and even over their shoulder, unlike most other birds that usually fly with their bills straight ahead.

The first group of three that came in back this evening circled widely, making a ten-mile circuit between the mountains, before again returning. Most make smaller circuits. Very few fly straight into the roost; most of the ravens are in no hurry to land. Thus, birds are coming and going continuously, and there is traffic in the air as at a large, busy airport where many planes were in holding patterns before landing.

When the western sky is already deep orange at the horizon and yellow above it, silhouetting the blue mountain ranges, a group of 15 ravens flies past me, heading west for Larkin Hill, just one mile away. There they stop, circle, then dive in pairs, and career in exuberant flight displays that last at least 20 minutes. I suppose they are playing in some updrafts, or they like zooming down the mountainside. The group of 15 gradually breaks up. One small band of them suddenly leaves the hill and flies further west in a straight line. Then, many minutes after these have left, another band leaves in the same direction, and only three birds remain to continue the aerial play. It is almost dark when these three break off, coming back toward me and the roost in the pines, into which they descend directly without further dallying. The others do not return, presumably traveling on to another, more distant roost. At dawn there were suddenly no ravens at the depleted baits near the cabin. The remaining birds had also left in the night.

Although I wasn't sure of the exact meaning of what I had seen, it fit a pattern that was consistent with what I and others had observed over the last eight years here. The birds would soon need another meat pile. Today had been a wonderful day for flying, and they had dispersed in all directions, alert for new meat, much in the same way that I walk in the woods for the fun of it, maybe to hunt deer. One or a small group of them may have found meat—perhaps the entrails of a deer—and returning to the roost here in the evening, they felt good and played in the air. Others joined in the fun. And when it was time to go to sleep, the thought of the new food pulled them west, and some of the others followed. To me, this is one of the most elegant systems of sharing that evolution has devised, because it does not depend on anyone's often unreliable altruism.

I watched the ravens returning to roost on many evenings, always wondering if Jack might be among them. Finally, on one evening later in the year, I was sure I saw him. Unlike ever before, a raven left its formation and fluttered all around me in the top of the spruce, as I stretched out my hand hopefully. I called, "Jack! Jack! Come here!" as it came within 15 feet of me, but the raven did not answer. It just looked me over closely, then flew into the roost in the pines to join the others. Yet several minutes later, it left the roost and flew around me again before rejoining its fellows.

THE MOON

Like everyone else, I have always known that sometimes the moon is full, then it wanes, becomes "new," and returns to full. Sometimes you see it in the daytime, and other times only at night. Seeing it regularly now, I'm for the first time trying to figure out its rough pattern. Of course, science has it all worked out in detail. But no matter, I don't have a clue. I want to discover it for myself, as our ancestors had.

My first surprise was to notice that the moon rises in the east. If the moon orbits the earth (which was the only fact I'd learned), then it does not seem at all obvious that it should consistently rise from that direction, since artificial satellites circle in all directions.

There is a full moon now, and it rises *just* as the sun goes down, appearing to replace it on the other side of the earth's horizon. I note that it travels across the heavens all night, and goes down in the west, just as the sun comes up again in the east. There is no sign of it in the heavens all day. I'm noticing also that when it *is* in the heavens during the day, it is always in a *crescent*. What, then, is the orbit of the moon round the earth versus the rotation of the earth itself? Both are occurring simultaneously. How can I distinguish them, since the sun also appears to orbit the earth, but in fact doesn't?

As a first approximation, the rise and fall of the moon on a schedule close to 24 hours probably is not a coincidence. It suggests that the moon must be in a relatively stable position day by day, while the earth spins inside the moon's orbit. Therefore, if the moon stays in a relatively stable position with respect to the sun, then when the moon is full here on the hill, it must be full also when seen from anyplace else on earth. This was another relevation for me.

Since we always see the same features of craters and "seas" on the moon, that means that it does not turn. But being necessarily *always* half illuminated and half in shadow, because it is round and illuminated from only the sun, it follows that the moon's orbit around the earth is in the same plane as the spin of the earth. Drawing a simple diagram, I could now see that if the orbit were around the poles instead, then we would see only a half moon at *all* times, and it would never change. Now, knowing the moon's approximate orbit, I could suddenly appreciate why an eclipse of the sun could occur *only* when the moon is in full shadow from the earth, and why an eclipse of the moon can occur only on a full moon.

Since the earth is rotating much faster than the moon is orbiting, the moon must rise and set on the same horizons as the sun. Now the question was, which way does the moon orbit: in the same direction as the earth's spin, or counter to the earth's spin?

If the moon had no orbit around the earth, then it would rise every 24 hours exactly. If it rises later every day, then it has moved away from, or in the same direction of, the earth's rotation and the earth must catch up. If, on the other hand, the axis of the moon's orbit is in the opposite direction to the earth's orbit, then it will have moved ahead—the two rotations add up, making the moon seem faster, and it will come over the horizon sooner. How much later or sooner?

If the moon goes once around the earth, or 360 degrees, in about 28 days, then the moon moves 12.9 degrees per day. Now we need to convert that figure to time. The earth turns 360 degrees every 24 hours, or 15 degrees per hour. So if 15 degrees is one hour, then 12.9 degrees converts to a fraction (.86) of an hour, or 51 minutes. So now I have a prediction! I can't wait to time the moonrise tonight and tomorrow, to check my prediction, to see if I'm right. If the moon rises 51 minutes earlier each day, then it's orbiting in the opposite direction to the earth. If 51 minutes later, then it's the same direction.

Without precise reference points and instruments, I could not time the moonrise to the exact minute. But after a couple of days watching the moon appear over the ridge, it became clear that it comes up nearly an hour *later* every night. So now I know that the earth's spin and the moon's orbit are not only in the same plane, but also in the same direction, at least approximately. To "discover" these facts was not like learning them from a book. They were *real*, and this process was a lot of fun. I wondered why my teacher in grammar school had not taken the whole class out onto the lawn to look at the moon and show us that the world is logical.

Astronomers and planetary scientists tell us that although the moon is an inert oversize rock, we may owe it our life on this earth. The moon was likely created by debris launched into space by the impact of a Mars-sized object with the earth. The collision caused the earth to spin, giving us day and night. We spin on an axis, or tilt, of about 23.5 degrees from the sun. This tilt causes our seasons. The pull of gravity by the moon, in turn, stabilizes the earth's tilt so that it does not oscillate and cause drastic temperature fluctuations. A one-degree change in tilt is thought to be enough to trigger an ice age, and without the moon's stabilizing influence, astronomers conclude that the tilt would have played havoc with the world's environments. Epochs with no seasons would have alternated with epochs of extreme seasonal variation, so that ecosystems wouldn't have been stable enough for advanced life, like the raven, to evolve.

HABITS AND HUNTS

Up to now, I have shaved almost every morning. It seemed like I needed some ritual to put structure in my life. Shaving was a start. But when you get up at 4 AM to be after the whitetail deer by dawn, your first thought is a cup of coffee. You are fumbling around in the dark because the feeble flame of the kerosene lamp on the table does not even illuminate the stove, where with a flashlight you check to see if you are hitting the coffee filter when you pour the hot water on the grounds. The stove is black, with only a few yellow flickering lines visible along the seams at the top. There is not much chance of finding a razor blade, much less using it in the appropriate way. And at night? Why shave after you get home to the pitch black cabin and repeat the fumbling around? After you've missed shaving for two or three days, another day without it doesn't change much.

Already I'm getting used to a stubbly beard. I'm now noticing for the first time that it's peppered with gray.

Eating, too, is changing. Aside from the morning cereal, I don't cook meals anymore. Like any animal, I eat when I'm hungry and when it's convenient. Perhaps the fire is hot, so I pop a couple of potatoes into the oven and have baked potatoes for supper. In the evening I might eat peanuts and drink a beer, or open a can of red beans. When I go out into the woods, I usually load up my pockets with raisins, an apple, or some crackers.

I can't think of a thing I'm wanting, except maybe a family to share what I have.

November 15
MOSS WATCH

Armed with a rifle, today I repeat my ritual of watching the sun go down from a perch up in a tree. There are gentle breezes this afternoon. As the wind blows through the ash tree near me, one that is laden with curtains of dry hanging tan seeds, it sounds incongruously like a wave washing over beach sand. Below me the crisp leaves rustle occasionally in a brief flurry, accompanied by high, soft peepings that remind me of tiny baby birds begging in the nest. It is a shrew. Having become attuned to them, I now hear them everywhere, every day. Why are their forays to and above the leaf surface accompanied by squeaks and peeps? You have to listen and to know, to hear the shrews at all. But the red squirrels' antics are hard to miss. Here, one scolds in a stuttering staccato of coughs and screeches that remind me of a two-cycle engine, barely running and badly in need of oil. The squirrel is thumping its hind feet up and down, keeping time with its vocal rhythm, not for musical reasons but for emphasis, no doubt. Then, suddenly, it erupts in a smooth "churring" that you might easily mistake for a miniature chain saw. Now it descends to the ground, hopping off through the dry leaves in a noisy series of leaps that, if you didn't see it, you'd think could be at least a doe, if not a small buck. The squirrel's pattern of noise on the leaves is varied, because it may travel by hopping over the ground from one tree to the next, it may dig for maple seeds, stopping to eat each one, or it may dash for a short ways by running.

From my perch in a tree, and walks to and from it I see the predominant gray and white tree trunks, and the kaleidoscopic pattern of browns from their fallen leaves. It is a rare event indeed to see a deer on a deer watch, but until I do I enjoy seeing the woods—particularly the stunning display of mosses and lichens that are now an open book. Before, when the canopy was green with leaves, and the ground was covered with herbs and small seedlings, these mosses did not stand out. Now they shine with luminescent brilliance.

Mosses are growing wherever they are not smothered by fallen leaves. All the old rotten stumps and decaying fallen trees are clothed in their green. The rocks protruding from the ground are covered with small cushions or thin layers of shaggy green moss, gray lichens, or both. In addition, the tree trunks are ringed with bright green growths at their bases. In the moister places, where there are no broadleaf trees whose falling leaves would smother these tough yet delicate plants, they also cover the ground.

Down near Alder Stream, as you come out of the dry tan swale grass and the gray speckled alders, and enter the tamaracks, you walk on a deep cushion of pale orange-tinged sphagnum moss. The moss is soft and spongy, but not resilient. The deep hoofprints of moose and deer are preserved here for weeks, maybe months. Sphagnum also covers the ground on the ridges, in the spruce forests, wherever moisture is retained by cachements of the underlying rocks. Where the ground is slightly drier, in the coniferous woods, there are smooth rounded hummocks of a pale turquoise moss. These hummocks vary in size from coins to basketballs.

Usually sphagnum moss (whose many particular species only experts can identify) and the smooth round hummocks both stand out from a distance, but all of the other green incrustations on stones, stumps, and logs blend together into a homogeneous growth. When getting on hands and knees to examine the mosses closely, however, you are ushered into another world of intricate design. There are distinct forms, each with its own beautiful shade of brilliant emerald, sap, and yellowish green. There are incrustations of moss with the texture of the smoothest, finest velvet, the color of sap green. In another velvety, furlike moss you can just barely see the "fibers," tipped in blond. Some rocks are incrusted with a burly, blackish peppercorn moss, tipped in light green. Others sport a wig of what could pass for close-cropped, wavy human hair.

The little green cushions and incrustations occur not only in association with trees and rocks. Where there is no blanket of leaves from above they also occur on the ground, intertwined with other plants to make a miniature green winter forest, which will, however, soon be covered with snow.

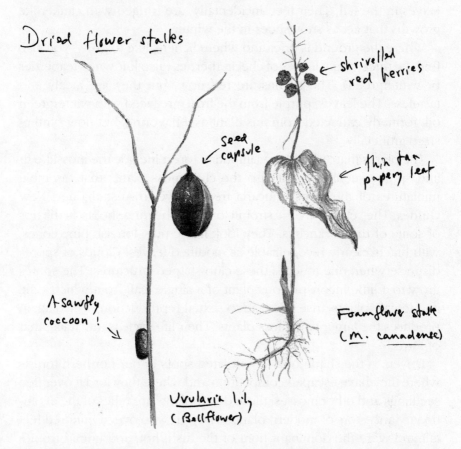

Dried flower stalks

← shrivelled red berries

Seed capsule

← thin tan papery leaf

A sawfly coccoon !

Foamflower stalk (m. canadense)

Uvularia lily (Bellflower)

A beautiful Lilliputian forest grows along with the moss in the few areas of the coniferous forest where there is a sufficient crack in the canopy to allow a bare minimum of light to reach the ground. Here among the moss now lie the plants that we saw in the spring with their white flowers. The bunchberry leaves are still alive, but have turned a rich purple. Their red berries are long gone, perhaps eaten by grouse, squirrels, or mice. The stalks of the foamflower plants are dry and wrinkled, but they still bear a few shriveled bright scarlet berries. There are the shiny trilobate leaves of the goldthread plants, also called canker-root. The yellow roots are said to make a bitter tea that stimulates the appetite and soothes cankers. I've never had need for either remedy. (However, partridges often have their crops full of goldthread leaves in the fall. Their feet, incidentally, are fringed with cankerlike growths that act as snowshoes in the winter.)

Where the ground is drier and where lichens like to grow, you also find the thick, waxy leaves of checkerberries (also known as teaberries or wintergreen). The berries are red now, but they are mealy and tasteless. The leaves, purple from the frost, are laced with wintergreen oil, formerly extracted from this plant as a flavoring, but now synthesized artificially.

The plants that "make" the Lilliputian forest include the mosslike fir and spruce seedlings, but also the club moss plants that resemble miniatures of the giant araucaria trees native to Australia and New Guinea. The "clubs" are the strobili, or spore-bearing shoots, at the tips of some of their branches. They look like crude, longish pine cones, with bracts clearly recognizable as modified leaves. Clouds of spores disperse when one touches these club-shaped structures. The spores grow to a little inconspicuous plant of a single leaf, from which male and female plants arise that lead to sexual reproduction, which again produces the larger club moss plants. Their life cycle is much like that of ferns.

Driven to the shade of a few isolated spots in our northern forests where they have escaped competition and where they lord it over tree seedlings and other mosses, these dwarf plants are relics of the evolutionary ancestors of modern plants. They grew to over a hundred feet tall and were the dominant flora of the lush, hot, and humid tropics some 250 million years ago. They are responsible for much of the coal that we burn today.

Whereas the mosses are restricted to their specific niches by competition and by moisture, the lichens are at least free of the restraint of moisture. They are epiphytes: they nourish themselves on the wind and the rain, and are free to occupy not only the bare open rocks where the sun broils down in the summer, but also the stems and limbs of trees. When the summer sun shines on them, they become hard and brittle, and enter a state of suspended animation. They survive freezing, thawing, dessication, and heat, but neither they nor the mosses survive being covered by the leaves of deciduous trees.

Like the mosses, the lichens grow in an amazing variety of forms. Examine a rock and you see a complex interdigitation of moss and a lime-colored lichen with short, columnar sporophytes. Another rock next to it may have a similar incrustation of different lichens, with small, cup-shaped depressions in the sporophytes. Another still has broad, dark green sporophytes without raised stalks, and still another has tall ones, with the cups flushed out in bright red, commonly called "British soldiers." Most of the lichens covering (or partially covering) every rock in these woods are of a flat type with lobes, and they are dull white or gray tinged with pale greenish blue. There are also rare patches of a bright orange lichen incrustation, and on some ledges you see colonies of the broad, flat, brown lichen called "rock tripe" that seems plastered on like small ragged patches of thin rubber.

Devoid of leaves, the deciduous trees look bare from a distance. But viewed from up close, you see that every tree is home to a vast assemblage of lichens. Here and there, especially on spruce and fir, you see tufts of finely branched usnea lichens that seem to sprout from one point on a tree limb and then reach in all directions. In contrast to these spottily occurring, pale green lichens, there are the more bluish-tinged lichens that grow in small lobes and that incrust the trunks and limbs almost everywhere that they can get a foothold. They do not attach themselves to fast-growing twigs with smooth bark, nor to tree trunks with peeling or shedding bark. From the top of my tree perch, I notice no lichens on the bark peeling off in strips from the hornbeam, but the branches and trunk above the peeling bark are covered with them.

The smooth-skinned trees host their own community of lichens, which grow in such thin sheets that you only see the color; to the naked eye they look like a layer of paint. Of these, the most conspicuous is a white lichen that gives all the smooth young maple trunks the appearance of having been splattered with round white dots. These dots come in all sizes and leave no tree in the forest unmarked by their distinctive patterns. (Curiously, however, I do not see these markings on my maples growing in the open field). Although most of the dots and splashes are a chalky white, there are also small round patches of a thicker "paint," with colors of pea green, light blue turquoise, and gray blue.

After many hours in my tree, with dusk approaching I suddenly hear what could have been a partridge. Yet, it has a slower beat. It stops, and I think, as usual, that it might be a deer. I peer into the direction of the sound, an alder thicket. I see nothing. I wait. Nothing stirs. Fifteen minutes go by, and the noise resumes, gently, slowly, and in my direction. Now I get my rifle ready and brace myself, making sure of my footing. Steadily the steps come on: "crunch, crunch, crunch, crunch . . ." And then I see a large brown animal.

It's a doe. Her nose is close to the ground. She has determined that the area is safe, and now she confidently walks toward the apple tree where I am perched about ten feet up. Walking directly beneath me, I hear her forcibly inhale as she sniffs the ground. She finds an apple and chews it loudly, then she lifts her head and scans all around. Now she resumes her sniffing and finds another apple, looks up at me, and we look each other in the eye. I see no alarm. She again bows her head and sniffs for more apples. I make a "psst" sound to get her attention. She again looks up at me, and then resumes her leisurely feeding. I am flabbergasted, having only seen deer as the shyest of animals. I feel no desire to kill this strategically disadvantaged animal, but I have an inspiration.

Gently I reach to the side and pluck an apple off the tree, then drop it. I hit the top of her head. Now I could boast about a perfect shot. I wonder how many other deer hunters will be able to brag of a similar feat.

Dusk is creeping on fast, and when I'm ready to leave my perch I move slightly. No response! I then start to climb down. The doe just stands and watches me, as if wondering what sort of a strange apparition could be up in a *tree*. A bear? Her predators—dogs, coyotes, and humans—are ground creatures, and apparently she only recognizes them in their usual context.

I now descend noisily. The deer stands still and watches. But as I take my very first step onto the ground she becomes a very different animal. As if electroshocked, she erupts in sudden, violent leaps and disappears with loud warning snorts.

November 25
CONNECTIONS

Last evening's rain turned into snow through the night, and this morning a world of brown decaying leaves, gray lichens, and bright green mosses has changed abruptly to a white fairyland. The wet snow has stuck onto every twig in a thick frosting. While the snow covers one world it now also reveals another.

A trail of mouse tracks leads from one woodpile to the next by the cabin. A snowshoe hare has hopped down the trail by the outhouse. And at the outhouse, there are more mouse tracks that look like hare tracks in miniature, but with a tail drag-mark in the middle. That makes it deer mice. Right next to the outhouse, a red squirrel has been on the stone wall, along which, going in and out of gaps in the stones, are the tracks of red-backed voles. A coyote passed an hour or so before dawn at the edge of the clearing. The new snow has only slightly blurred the sharpness of its tracks.

It is the kind of day I've been dreaming of, a perfect day for deer hunting. I can now walk quietly except for a low, muffled crunching under my footsteps. I can now go through the woods in virtual silence, reading in the record of the snow when and where the deer have passed, how fast they were traveling, and where they might be going. It is as if I've suddenly gained a secret insight into some dynamic mechanism, and my excitement mounts with the new power that this knowledge provides.

I hesitate at the cabin door, stopping to watch the ravens flying through the gently falling snowflakes, setting their wings and gliding down to the dead calf I've left them as an offering. They feed, then frolic like kids by rolling on their sides and backs in the new fluffy snow, often grabbing a stick with their feet and playing with it in their bills as they cavort.

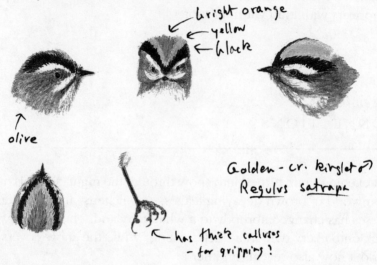

It has been a very successful deer season, I decide, even though, as also in most other years, I have not fired a single shot. For almost one month now I've been out in the woods every day for two to ten hours. I've hiked probably a hundred miles, exploring Potter Hill from the sphagnum-fir bogs up to the oak-beech ridges. I've become tired, and I've sat still for hours on end in absolute solitude, with all of my senses totally engaged. I've heard red squirrels scampering over dry leaves from over a hundred yards away, and I've seen them hold and chew a spruce cone three feet from me. I've had a golden-crowned kinglet land within two feet of me, allowing me to look into its tiny black eyes and see the orange flame of feathers framed in yellow, white, and black on its fluffed head. I've heard a pileated woodpecker drum exuberantly, and I've been witness to a fisher killing a porcupine. I've spotted the shiny white of a hare against the moist, dark leaves at dusk. I've seen strange and wonderful mosses and lichens.

It is true, of course, that I need not have carried a 30-30 lever-action Winchester rifle with open sights in order to wander and explore as I have done. But that is not the relevant point. The relevant question is: *Would* I have? Most of the rewards we get in life come to us obliquely.

Satisfied with this year's hunt, for once I stay home in the cozy cabin, drink coffee, and read the paper.

November 28
MORE FLIES

With winter coming on, the cabin is becoming an ever more popular retreat. A new crop of mice has already moved in. They keep me awake at night as they play tag and tumble and fight in the crawl space of the ceiling above my head. As far as they are concerned, I had thoughtfully provided them convenient lodging when I built the cabin. Actually, the *intended* space was for insulation, so it also contains sheets of styrofoam that they crunch on incessantly. The small white styrofoam flakes rain down like light snow through the cracks. Unlike snow, however, these flakes do not melt and go away, and when you try to sweep them they only roll and bounce.

The big black cluster flies come out of the same cracks at night if they see my light next to the bed. Some come buzzing drunkenly off the ceiling, motor around loudly, and butt against the light. Then they crawl under my sheets. Ideal bed companions they are not.

If I make the cabin warm and cozy in the daytime, they also come out in droves. I thought I'd killed them off once already, but not so. In the last few days on more than one occasion, there were enough in their collective hordes to make the windows hiss. Well, for a while. I'll propose a new ad for the Dustbuster: "Scientist used Dustbuster to suck 12,800 flies out of his house! You, too, should own a Dustbuster." Or maybe I should submit a new category to *Ripley's Believe It or Not*. "What is the all-time record number of flies that someone tolerated in one room?" I don't know. But if you are wondering how much 12,800 cluster flies comes to, it's nine and a half cups full, level.

If I were to see a fly for the first time, it would be a marvel of creation. There certainly is no danger of eradicating them, however. I've been destroying flies in my cabin for years, and there were never any fewer the next year.

December 3

The snow has melted and the ground is bare once more. Two weeks ago, the unprotected earth froze solid. Now it has thawed again, and there is a huge difference. On any walk in the woods before the heavy frosts, you heard the high-pitched twitter of shrews just in and under the leaf mold. Lately I've heard not a one. They probably froze to death by the millions. But it only takes a few survivors to repopulate, and there are survivors. They are likely those with some idiosyncratic trait—perhaps a predilection to hole up in huge rotten stumps, or to burrow more than their fellows—which the next generation will inherit. Next year these "burrowers" might in turn be at a disadvantage, if there is snow cover and they are then wasting their efforts.

The lack of snow also affects the snowshoe hares, known in Maine as "rabbits." With their white winter coats, they (as well as the weasels) now stand out in the dark brown forest as if they had been painted in neon. The signal for their coat color is given by the changing hours of light and darkness. Since they are white now, it could mean that through evolutionary time on the average, there has been snow this time of year. We are so far having an unusually snow-free winter.

The red squirrels, hiding in the branches of conifers, have also changed their pelage. It is much thicker than in summer, and it now has a brilliant russet hue that seems as much an ornament as a protection from the bitter cold. Last winter, when the balsam firs yielded a massive harvest of cones, the red squirrels fed on their seeds, as did huge flocks of pine siskins, purple finches, and evening grosbeaks. This year the cones are absent; instead, there is a massive crop of ash seeds, which were absent last winter. The ash seeds will probably again attract finch flocks, but I have not seen the squirrels feeding on them. Instead, under many of the trees with fungi, there are piles of little white chips that the squirrels have dropped while feeding on them. Perhaps these are the peelings as they eat the more tender meat inside. Often, too, I come upon a frozen apple or an acorn stashed in the crotch of a tree, far from any apple tree or oak. These, too, are the works of red squirrels, who will undoubtedly retrieve their caches after the ground is under several feet of snow.

The hunting season is over, and I now have no real "reason" to spend my days in the woods anymore. I need another project.

December 4

My editor came for a visit, and after I was alone again I felt a tugging at my heart—I needed to see Stuart and Erica.

It is about 11 PM when I get to Burlington and check into a motel. The trip west to Vermont over the mountains has been a five-hour drive. For $45, I luxuriate at midnight in a tub brimming with hot water. I have not missed this treat, because it was not in the realm of possibility. However, I now enjoy it immensely.

I sleep soundly and get up early, to have a "Continental breakfast" of a blueberry muffin and coffee. The muffin is tasteless, being a very poor substitute for bread, which *is* served on the Continent. The tropical plants in the lobby, I notice, are fake as well. As with the muffins, people get used to imitations—false plants that you don't have to water, false animals that the kids can just throw in a corner when they don't need them. You only cuddle them when you feel like it, and they never make a mess.

Christmas is nearing. Erica will go to California to see her mom; Stuart and his mom will drive to Michigan to see his grandmother. My mother always had a Christmas tree decorated with cotton, tinsel, colorful globes, and real red candles that were lit when the family came in and saw the tree for the first time on Christmas Eve. *Green* trees that have been picked out of the woods the day before don't ignite from a candle flame. Singed needles only add to the celebration because they crackle like sparklers and give off the pungent aroma of the evergreen woods. As the candles burned down, we sang Christmas carols, and then Papa passed out the presents. A day or so after Christmas, the tree was discarded. That way, Christmas Eve remained special.

Lately I have grown tired of singing Christmas carols among grown-ups. I don't care to go through rituals of card-sending and present-giving. I think of a starving boy I read about in the newspaper. There are kids like him all over the world, needing so much. I remember too being nine and receiving a Care package from the States—a yellow pencil, a peanut. These things seemed like wonders. And then I have an idea: maybe I can invite some disadvantaged kids up to my hill on Christmas Eve. We can set up a Christmas tree outdoors, build a great fire and roast marshmallows, and have a Santa Claus stride out of the forest with presents . . .

December 8

I sat in the third-grade class at the Hinesburg Elementary School, and when it came time to show my slides, Stuart stood up beside me, held my hand, and said to all, "This is my Dad, Bernd Heinrich. He lives in a cabin in Maine and studies animals . . ." And then he told them what animals I studied and what I'd talk about, and how he, too, had lived at the cabin.

We looked at slides of frogs, owls, and lizards, and Stuart then showed them the "awesome" beetles and other insects that he had picked out from my collection to bring in. Everyone rushed up and crowded closely around. The students asked questions, made comments, and said they wanted me to stay all day. When I finally had to leave near noon, they asked when I'd be back. Stuart wanted to know if he could be an assistant on my raven project, and if he could go in the woods with me to look for beetles.

Erica may be less inclined to think about searching for beetles in the woods but she, too, shows aptitude for trying to puzzle out how the world works. She worked in the summer at a biotechnology firm. She shows me her notebook full of Western Blots, gels, chromatograms, and DNA fragments "chewed up" with enzymes, matched with probes, and it all seems just as wild as a tangled forest.

December 9
THE LUNAR ECLIPSE

On my return trip to Maine, I collect some raven food from the local dairy farms—nine dead calves. The cold weather is not kind to young animals. I always find a plentiful enough supply to load up the back of the truck—sufficiently to draw a stare or two, and to please the birds.

It is 26°F inside the cabin. I suppose the sun has warmed the house, because the nighttime temperatures have been twenty or thirty degrees lower. The sink drain is useless now, because water freezes in it instantly. The water in my two plastic carboys has been frozen into solid blocks.

It is a clear day, which should be ideal for watching the lunar eclipse that is expected tonight. I've gone to town to get steak, and I've left a note for Ron and Syndi to come up and join me by the fire to watch the moon.

Official sunset is at exactly 4 P.M., and I watch the full moon simultaneously rising in the east over the top of Wilder Hill, on which are silhouetted the ragged tops of red spruces, like little black teeth. Through my 10×40 Leitz binoculars, normally trained on black ravens, I now behold the moon's luminous sphere, clearly distinguishing the dark lunar "seas" and the pockmarked surface of giant craters. Just two days ago, the one- to two-mile-wide asteroid Toutatis (discovered only in 1989), whizzed past our own planet at 85,000 miles per hour, missing us by just 2.2 million miles. Astronomers called this a "hair-thin margin of safety in celestial terms."

If Toutatis had hit the earth, it would have created a crater 30 miles across and thrown up enough debris to shut out the sunlight. The plants would have died, and almost everything else along with them. Were this asteroid to hit an ocean, it would create massive tidal waves that would roar far inland on all continents. The astronomers say we can expect Toutatis to visit every four years, coming within one million miles (the narrowest limit of their calculations) in 2004.

There are millions of other large objects out there, some of which have in the past been swept up in collisions with the moon and the earth. Luckily, most bodies that are less than a kilometer in diameter pose no threat, as they would burn up in our atmosphere, so we do not track them. This month, comet Swift-Tuttle is returning after 130 years, and should be visible with binoculars 30 degrees above the horizon in the southwestern skies near the tail of the constellation Aguila, the Eagle. This comet was responsible for August's Perseid meteor shower. Tonight, however, I'm only going to witness the collision of earth's shadow with the moon.

I build a fire in the stone fireplace next to the cabin. As the moon rises in the east, the orange flames of my fire shoot straight up, carrying bright sparks into the cloudless, darkening sky. I sit on a log next to the fire, trying to warm my feet. Temperatures are rapidly dropping to 0°F, so I am fortunate that it is not windy. There are no sounds except the flickering of the flame and the hiss of some burning wood that has not thoroughly dried. The moon becomes ever brighter as the sky darkens, and it climbs above the ridge. Its radiant white light glares off of the snow around me.

By 5 PM the sky is nearly black except for the bright illuminating moonlight, and I see the first darkening of the moon at its lower margin. The shadow of the earth is now creeping across the face. Slowly, ever so slowly, the shadow consumes the entire moon from the bottom, gently creeping upward. It is a strange sight indeed to see a crescent form by darkening it at the bottom (from the earth's shadow) instead of at the sides (from the moon's own shadow). For thousands of years, this strange sight has frightened millions, and has been thought to portend evil or calamitous events.

In an hour, the bright moon seems to have been gobbled up completely. Only a thin bright crescent remains at the top, and that, too, disappears in another seven minutes. For the next hour and a quarter, the moon no longer gives off its bright reflected light, yet it remains faintly visible as a very pale copper sphere, due to the light striking it that has been bent and weakened by going through the earth's atmosphere before being reflected back.

The virtual disappearance of the moon glow then lights up the stars in the darkened sky. As I look at the moon with my binoculars, I also look past it into our Milky Way galaxy that blazes a wide swath from east to west, directly in the path of the moon.

I have never gazed with binoculars into the Milky Way before; the millions of stars visible with the naked eye have always been impressive enough. But what I see tonight with ten times magnification is almost frightening. Some mysteries are a little too daunting for an average weekday.

At 7:15 PM a bright glow reappears on the lower left surface of the moon, and as the light returns, a coyote howls briefly. Mars rises, luminous and red, just to the north of the moon, and as a bright crescent of light spreads over the moon's left side, the myriad stars recede. By 8:30 PM, three and a half hours after the show began, the moon is almost overhead and again shines, fully. As the sun's light, reflected by the moon, reflects further off the brilliant snow, the woods become bathed in a silvery glow, and I see the gray shadows of the trees.

I had predicted that an eclipse could only occur on a *full* moon, and that it must start at the bottom and move up. I am thrilled now to see that this is exactly what happened. So why isn't there a lunar eclipse *every* month? When at last I allowed myself to research the subject, I learned that the moon's orbit around the earth and the earth's orbit round the sun are not on *exactly* the same plane—they are tilted at a five-degree angle with respect to each other. The moon's shadow therefore misses the earth during most new moons (potential solar eclipses), and most full moons (potential lunar eclipses) pass above or below the earth's shadow.

December 11
WIND

The wind is howling in from the north. It is dark and cold, and tiny snowflakes are swirling through the trees. There is a bite to the air. I sense a storm coming, but I cannot be sure. It is certain, though, that something is afoot, and it animates me like the descending dusk animates the wolves to start the hunt.

Even the ravens seem restless. They came only briefly to feed at about 8 AM, and then they left, first playing in the wind. They threw themselves at the gusts above the trees to be yanked skyward, and then they tumbled back down in twos and threes. As they were feeding on the calves, I could see with my binoculars that their black feathers were tinged with white frost, as a man's beard and fur collar might be from exhaling moisture into very cold air. Not only were the ravens frosted around their black bills, but also on their breast and back feathers, and on the forward leading edges of their wings.

A fierce restlessness draws me out into the woods. As an excuse, to give this urge legitimacy, I decide to search for golden-crowned king-lets. Are these tiny wraiths still here after the sub-zero temperatures we've had? I suspect they are, but I must *see* them to be reassured. I wonder how they can stay alive in these woods, all winter long, on a diet of insects gleaned from open branches. Very few people know the birds exist here at all, so I feel like I'm delving into a deep secret. If I can find out how they do it, then I'll know something that nobody in the whole world knows, and that prospect excites me. I do not yet want to form a hypothesis to test, because as soon as you make a hypothesis, you become prejudiced. Your mind slides into a groove, and once it is in that groove, has difficulty noticing anything outside of it. During this time, my senses must be sharp; that is the main thing—to be sharp, yet open.

Before going on a hunting expedition, some Indians of Amazonia make themselves feel godlike, with sharpened senses and increased strength, by "taking frog." Their hunter's magic is a potion scraped from the skin of a green frog, *Phyllomedusus bicolor*. I sharpen my own senses and enhance my feeling of well-being by imbibing a brown brew that is liberally served at the Farmington Diner, made from pouring hot water through the ground-up seeds of a tropical shrub. I enhance the flavor by adding a dollop of the udder secretions of a cow, plus the crystallate of the juices of the cane plant. This morning I again have this potent concoction in some abundance, to which I add a "Hunter's Breakfast": a stack of four pancakes, two eggs over easy, four slices of toast with jam, four strips of bacon, and a donut for dessert.

When I get back into the woods to hunt kinglets it has snowed lightly, so I can see yesterday's animal tracks and distinguish the fresh ones from those made earlier today. In addition, the snow cover is thin enough that the small animals cannot yet burrow under it, so I have the advantage of seeing their tracks from one patch of ground cover to another. I find tiny half tunnels, where shrews tried to keep covered by snow, that look like they could have been made by beetles. Here and there I see little mouse highways leading from one log to another. Most of these are the tracks of low-slung voles whose bellies drag in the snow. I see fresh red squirrel and hare tracks, but curiously none at all of weasel.

As I leave the cabin clearing, I soon meet up with a troupe of over ten chickadees. Golden-crowned kinglets often accompany the chickadees, so I follow this troupe to look for them. After half an hour, however, I am convinced that these chickadees are unaccompanied by any other birds. I leave them on the hill and descend to Alder Stream, to search among the balsam firs.

On the way down I see a red squirrel sitting hunched up and immobile, on a maple tree. All around under this tree, the leaves have been recently turned over. The squirrel apparently has been harvesting the fallen maple seeds. The squirrel is silent, unlike in the fall; they now all seem to have lost their voices.

There is not a single fir-cone bract on the snow, nor can I see a single fir cone on the trees. A major food source of squirrels and many finches is totally absent this year. At this spot last winter, I saw big flocks of brilliant yellow-marked evening grosbeaks that are the northern equivalent of parrot flocks. There were myriad purple finches, goldfinches, red polls, and pine siskins. Five to seven years ago, there were also two species of crossbills, but there have been none of these since then. Some winter, they will return, but so far this winter, the woods have been almost devoid of them as well as almost all the lively, colorful finches. I miss them.

Through the wind, I'm straining to hear the high-pitched calls of kinglets. Only after about an hour am I rewarded by the thin, squeaky calls of these birds that are as unobtrusive as the breeze, and just as easily not heard. I only make out one lone bird at first as it hops incessantly among the thick branches of a balsam fir, occasionally hovering like a hummingbird. It is not much larger than one, and it is practically invisible in the fir's big limbs. I soon hear another bird, possibly two. No others accompany them. I have learned something already: kinglets may travel in very small troupes, and they may travel independently of other birds. Within four or five minutes the birds slip from my sight and hearing, having been at the bare lip of reality to begin with.

I have not picked a good day to search for kinglets, because the moaning of the wind and the creaking of the trees so thoroughly mask their delicate sounds that these birds, which can never be followed very far by vision alone, cannot be followed at all. But whenever I have an idea, I need to act on it as soon as possible. Conditions are seldom ideal, and if one waits long enough for ideal conditions, then one is just making excuses.

The windstorm picks up at night, and I listen from my snug nest just under the roof at the north end of the cabin. From there you do not hear the rustling of the few remaining dry beech leaves on young trees. You do not hear the twigs snapping, nor the trees creaking against each other. But you do hear the uneven rhythm of the wind's roar. Unlike the surf, this rhythm is infinitely varied; there are no even waves of sound. There are seconds of silence, then a soft, high hiss, perhaps followed by a deep, dull roar. The roar may die down in one direction, then pick up in another. Suddenly the cabin shakes. Staccato gusts come and go, as do long, gently varying moans with sharp whistles. Suddenly all is quiet, and you wonder when the next violent blast will rock the cabin. The infinite variety of the wind swirling over and through the forest is my evening's listening pleasure, and I find it soothing.

This cabin is built with heavy, solid logs. I feel secure inside. The cracks between the logs are chinked by oakum that I have laboriously pounded in. The woodstove is throwing off heat, and I bask in its glow.

December 12
RUMP STEAK AND ROADKILLS

I did not get "my" deer this year, but perhaps someone took pity on me, because he or she left about a dozen pieces of meat wrapped in white butcher-paper in my pickup truck down by the road. There were pieces stamped "roast," "rump steak," and various other choice cuts. The generous giver did not even leave a note. But, with or without a message, I've been gratefully eating it over the last two weeks, and pronouncing it prime. All the time, though, I've been wondering where it came from. Today I saw Bill Adams in town.

"Did your ravens like the meat?" he asked.

"What meat?"

"The meat I left in your truck."

"You mean the steaks?"

"Yeah—I got them from a friend. She was cleaning all the old stuff out of her freezer to throw away, and I told her your ravens would love that stuff."

"You should have left a note."

"Yeah, you're right . . . How did they like it?"

"I'll have to ask," I said, feeling a slight wave of nausea.

Although my taste buds told me it was fine, I subsequently *didn't* eat the rest of that meat. I don't know why, but I suspect it's the same reason I stopped eating roadkills after my college roommates made strange noises in the bathroom, following their enlightenment after meals I had made from run-over muskrat and raccoon.

The roads are a great slaughtering ground for all animals—why not make maximum use of them? On my jogs along the highway in spring and summer, I have seen enough songbird carcasses to stock a museum, and while it troubles me that all these birds are killed, it troubles me immeasurably more that it is unlawful to make use of them—that you are supposed to let them rot. Most states have strict laws against using roadkills. In Oregon, removal of a roadkill can carry a maximum fine of $2,500 and a year in jail. In the town of John Day, Oregon, folks recently staged a big, free Saturday dinner to celebrate the acquittal of one John Taylor, who had been tried for the crime of butchering a roadkill deer and distributing the meat to poor folks. Topping the dinner menu were roadkill rabbit and squirrel, and of course also deer.

December 13

The wind continued another night, sounding at times like a speeding freight train. It continues this morning. I hear moans, shrieks, rumbling, and clatter. I feel the storm's breath and my insignificance in its power, as if propelled by it like one of millions of leaves. It feels like freedom: I can go with the flow.

I have the luxury to explore where the wind blows. Yesterday I found the handsome purple buds on an elderberry (*Sambucus*) bush, which I drew in watercolor. I like the way they look. Maybe I should draw those of other forest trees and bushes to go with these. And so I went on a bud hunt, gathering many varied budded twigs. Today I will paint some of them, thinking about the leaves and flowers of next spring's growth packaged fully formed already when the old leaves

Beaked hazel

Red oak

Elderberry

Skunk currant

Red maple

Hobblebush

Sugar maple

Some winter Woodland buds

were shed two months ago. The buds are of exquisite design, but to see them truly I have to draw them, and to draw them I have to see them.

I started off by drawing the twigs and winter buds that I collected yesterday. By 2 PM I had completed a picture of 11 of them, and felt a reasonable sense of accomplishment. Now, with something concrete to show for the day, I celebrated by finally doing the dishes. The last time I had done them—I do not remember for sure, but I know it was at least two weeks ago—I had washed all the dirty ones except my coffee mug, because I planned to "dirty" it immediately afterward with more coffee. While the kitchen stove was humming to heat the water, I also threw three potatoes to bake in the oven, and put onions, tomatoes, and cabbage on to stir-fry. Meanwhile, I took a shower outside, where it was 34°F, but I didn't feel uncomfortable. By 3:15 PM I was done with dinner, dishes, and my shower; I was having fresh coffee and feeling free again.

December 15
YOU CAN NEVER GO BACK

The Hinckley Good Will School is only an hour away on Route 2, the main thoroughfare that runs diagonally west and northeast, bisecting the entire state. It is a two-lane blacktop highway, with a little variety store at every crossroad. Most of these establishments are decorated with bright red Coca-Cola signs, Marlboro posters, and/or neon "Bud" signs. As you come into a little town, you see signs along the road for the Church of the Nazarene, the Pentecostal Church, or the Union Baptist Church, and seeing few signs for other public places I think of Edward Abbey again: "Any town with more churches than bars has a serious social problem."

The church buildings are somewhat less grand than some of the houses at the edge of town that are surrounded by gnarled sugar maple trees. These older houses are usually painted white, and the barn connected to them is almost invariably rust red with white trim. Driving by, I glimpse a shaggy horse standing stolidly by the barn door next to a manure pile . . . fields covered with corn stubble and surrounded by alder thickets . . . raspberry and spirea brambles . . . stray apple trees with now frozen brown apples still on the branches.

There are also new little houses painted in bright colors—turquoise, yellow, light blue. Often quite near them are the remnants of the old farmhouse where the grandparents perhaps still live. Its graying clapboards are loose and falling off, and plastic sheeting has been nailed over the windows. Rusted hay rakes; manure spreaders; remains of trucks and cars, washing machines, and other unidentifiable cast-off paraphernalia are strewn about among dry stalks of burdock and goldenrod.

Between Mercer and Norridgewock is a long stretch of woods. Logging roads go off to both sides, and there are many small recent clearings. In each little clearing stands a modern, avant-garde type of dwelling—the trailer. Some of these have a lawn near them, and you see a broken lawn chair or two left outside from last summer. But on almost every dwelling there is already a fresh green wreath of balsam fir with a bright red bow, as well as strings of Christmas lights and plastic Santas, illuminated from within by a lightbulb, that are plugged in just after Thanksgiving and unplugged long after New Year's.

Between Norridgewock and Skowhegan, and then all the way to Hinckley, the big Kennebec River runs along the left side of the road. It is frozen over and looks smooth and black, like burnished steel. The school lies along the river and I approach it with a mixture of sadness, trepidation, and nostalgia.

I turned off toward the river on a dirt track opposite Averill High and went over the railroad crossing onto the field where for many summers I had toiled in the garden on the river bank along with many other little boys. I wondered if the kingfisher still nested in the sandy banks there.

The ice on the river was smooth and transparent, not crisscrossed with the white etchings of skaters' tracks. I thought of the Saturday afternoon outings when we had held bonfires on the banks and the kids had gone skating.

The river was strangely quiet today. I did not even hear the loud "poings" travel up and down it like dull muffled thunder, accompanying the white cracks in the ice like horizontal lightning. From my bed in Pike Cottage, I used to hear the river talk at night, and I had dreamed of the Far North.

Prescott, the administration building, looked the same, except that there were a dozen cars parked outside. Inside, I stopped at the first office that I came to, lest my courage leave me, to see the headmaster. This was a first, coming here without having been sent for a licking or a lecture. But the blond secretary acted more like she was trying to prevent a meeting than to facilitate one. The headmaster was in conference, I was told, and would not be available. She told me I'd need to see some-

body else, with some other title, who'd handle "activities." So I went
where directed and talked to another secretary. I told her my dream of
giving some kids a Christmas. She smiled. "A nice gesture," she offered.
She also recognized my name. "Oh, you're the famous runner."

But could I see her boss?

"He usually is here by eight, and sometimes earlier. But I haven't
seen him today." She didn't know where he was, and was not willing to
call his home. But she did offer the formality of taking my phone
number so he could call me later in the week. I told her I'd take a walk
and come back a little later.

I went up "Uncle Ed's" Road in back of Guilford Cottage. This had
been one of my favorite spots, where the old sugar maple tree had a
rope to climb and swing from. The tree was still there, but the rope was
gone. When I was in the seventh grade, I had gotten a set of eggs from
the red-breasted nuthatch whose nest hole was hammered into a
rotten limb on the same tree. I had punched holes in the ends of the
tiny white brown-spotted eggs, blown out the contents, and hidden my
illicit treasure under the floorboards of an abandoned building. I once
had found a baby raccoon in another hollow maple nearby, and in the
spring the yellow and blue violets had blossomed underneath.

Fresh logging signs were everywhere. There were skidder drag-
marks, knocked-over trees, slash, and stumps. The contrast between
what my mind expected and what it saw jarred me deeply, only
because I was comparing this with my dreams. Others would have
seen a reasonable logging operation, that left a reasonable forest intact.
My dream of giving the kids a Christmas would also remain but a
dream. This was a prep school now. The kids had a social director.

December 19

At 3:30 PM, I was back from town, rounding the corner of Hills Pond
that was now thickly frozen over under smooth, thick black ice. Two
ravens came languidly flying down the road, in tandem. They were so

close to me that I could see them flicking their heads from side to side as they flew along. After I stopped by my mailbox, I heard the low, sharp rumble of an ice crack shooting across Webb Lake, three and a half miles away. The sky was overcast, and I hoped for snow.

The evening was moonless. As it got dark, I watched Venus and then the much less bright Saturn appear in the southwestern sky. I built an outdoor fire and sat and observed the heavens. Orion came up in the east, Venus and Saturn went down, Mars came up in the east, and the Big Dipper rose in the north above the black silhouettes of the spruces. The Milky Way cast a bright swath across the heavens. I felt content next to my fire, and then, despite the cold, I drank an icy beer. It enhanced my feeling of well-being and I felt in the mood to party. The winter solstice was only three days away, and the urge to have a raucous good time among friends took hold. I jogged down the trail to the telephone, to call up reinforcements to engage in pagan rites under the stars. Nobody happened to be in. I trudged back up to the cabin and reluctantly went to bed.

It was a sunny day and, out of the blue, a raven flew near, circled closely around me, then landed at my feet. He bowed his head slightly, fluffing his feathers, and made soft cooing-begging noises. It was Jack. His greeting calls showed how glad he was to see me. I sat down and he hopped onto me, showing affection. I wrapped him in my shirt and held him to my chest. He became quiet like a puppy. At that instant I realized something was wrong. Jack would never allow himself to be wrapped up and held like that. I opened my eyes and through the cabin window saw the sliver of the moon just over the horizon. I rolled over, feeling my body rub the blankets, and as I again grabbed onto the world with my senses, the dream gradually faded.

WINTER

December 21

_____

The water is now frozen over even deep down in the well. I have to break the ice with a long pole before I can lower a bucket into water. But I think (again) that I've finally conquered the mice. I caught the last one of the latest crowd two days ago. (Famous last words.) Since then my butter no longer has teeth marks on it each morning. There are no more styrofoam chips floating down from the ceiling. No more scurrying and gnawing to disturb me each night. Peace at last. The flies have been somewhat less numerous the last few days, too, although a delicate green lacewing has just landed on this very page, after making a circle or two around the kerosene lamp. Lacewings never are numerous, and they don't buzz when they fly. I don't mind sharing the cabin with a few of them. Normally at this time of year, I find them under loose bark on dead trees, where they hibernate for six months or more.

The red squirrels are not deterred by the cold. The one that lives along the path to the mailbox seems to thrive on cached apples. I've found 15 apples in all. Some were up to 11 paces away from the original tree, where deer and other animals have by now removed all the apples from the ground. I wonder: does the squirrel *remember* where it has cached all of these apples? Is it possible that it finds them at random, as I just did? There is a simple test, and I run back to the cabin to embark on this project instantly. Next to the cabin is an apple tree where the wildlife has spared the apples, and I gather 15 of them. With a paper pad and pencil, I now draw a map showing where all of the squirrel-cached apples are, and I cache 15 of my own in similar places, such as in forks on exposed branches. It will be fun to come back in a month or two, to see which apples the squirrel has retrieved. If as many apples that *I* cached are gone as those that *it* cached, then the squirrel probably does not retrieve apples by consulting its memory. (I checked just over two weeks later—*all* 30 apples were gone!)

December 23
RAVENS

For the first time the snow is drifting down steadily in silent white curtains. Not a breath of air disturbs the tranquil spectacle. It seems almost eerie that something so stunning can be so silent.

The ravens are in their element. They are masters of the cold and snow. Those in my aviary (recently captured in a wire trap in the woods) frolic by rolling on their backs like playful puppies, as they always do to enjoy a new snow.

To lure these "wild" ravens closer to me, I crushed some peanuts into small bits so it would take longer for the birds to pick them up (otherwise they grab the nuts and hop or fly off), and stood and talked to them. I don't know if my talking helped, but it certainly didn't frighten them. They needed to be reassured that I wasn't doing something surreptitious to lure them close to my feet. Ravens are extremely perceptive of intentions. One of the boldest ones (marked with a *K* on her wing) managed to pick up enough small pieces to cache. She held her loot in her closed bill: not a bit of the tiny peanut fragments was visible. But as she nonchalantly walked off to hide what she had collected, all the other ravens' eyes were fixed on her. And when she stopped at the edge of a big rock to cache her peanuts, several birds left their perches, swooping toward her before she had opened her bill, as though knowing what she was about to do. But she knew what *they* were up to, too, so she did not open her bill, but flew away instantly when they came. The others gave a brief chase. I marveled how they *knew*—I could only see a slight bulge in her throat area, one that someone not familiar with these birds would never notice. Is *this* what the others saw, or was it her suspicious behavior? Whatever it was, they knew her intentions, and she knew theirs.

With my ravens I wear two hats, literally. Today, I wore the red hat, the "good" hat. I wear that when I bring them food. When I wear the black hat that pulls down over my eyes, they fly away in fright. That's the hat I always wear when I catch them in the smelt net to weigh them. (I am doing an experiment examining the effect of dominance on ability to get food.) It is important to give them clear signals, if one wants to stay on their good side.

Being on their good side is fun. I can sit there for hours and watch them, and I intend to tame them in the same way that they are tame towards other large animals, such as the wolves, bears, and coyotes that ravens normally meet at carcasses. Normally they fly off when you look at them from a mile away—hat or no hat. Trust is much harder for them to learn than fear, because of the disparate risk. With the one the bird is risking its life, with the other only a meal.

December 25

Yesterday in the late afternoon I went to my mother's house in nearby Dryden, looking forward to a hot bath as much as to the Christmas celebration. My sister Marianne was there with her husband Charles and their two grown sons and their girlfriends, along with the Kims, a Korean-American family we have "adopted" (or perhaps they adopted us through the friendship from high school, college, and beyond, of my nephews Charlie and Chris). In all, there were ten of us. Mamusha had outdone herself and cut the Christmas tree herself in the woods that morning, and she had prepared a meal featuring chicken, roast goose, mashed potatoes, and red cabbage.

We didn't sing Christmas carols, and I felt thankful for that. We had plenty of Christmas cheer, and talked about our separate lives in Florida, North Carolina, Vermont, New York City and Ithaca, and West Virginia. I appreciated the ice-fishing equipment that my nephews gave me, and Marianne and Mamusha were both very pleased with the framed and matted photographs that I had taken of two of my grand-mother's paintings that we located in Poland (but were unable to take out).

By 8 PM I was walking back up the path to the cabin. The wind was howling and the temperature was dropping fast. I built a fire in the woodstove, heated some tea, and browsed in a magazine.

I didn't sleep well, mostly because the cabin was rocked by gusts of sub-zero winds. I kept pondering how kinglets, no larger than the end of my finger, could survive in this cold. You have to *feel* this cold, to absorb its power. I was huddled, shivering, under thick covers, from which I ventured at intervals to put more wood in the stove.

29 December
WINTER ECOLOGY

The students arrived last night. By ones and in small groups, they came and made their way up the hill. I had a huge fire blazing in the fire pit in front of the cabin, and we banded around it, sharing some firewater.

Every winter during January (the semester break), I invite to the cabin a select group of a dozen students who have signed up for Winter Ecology. This isn't three lectures per week, Monday, Wednes-day, and Friday at 10 AM. This is all day, every day, for 15 days. I tell them from the outset that this will be a challenge not only in terms of learning, but also in terms of getting along and coping under cold and primitive conditions. Invariably a group spirit develops, and we part with the feeling of having had an "experience."

This morning I got up as it was getting light, and my activity around the stove and the smell of fresh coffee quickly roused them all out of their sleeping bags. I would start them off on the first morning with a hike to observe animal tracks.

We took off after a breakfast of cooked oatmeal. The conditions were perfect, because the three-day-old snow layer would have recorded animal activity over a number of days, so you could compare old tracks and new. The snow layer was thin and slightly sticky so the tracks were distinct rather than immediately filled in as made. Before we trudged back in the mid-afternoon we had seen the tracks of coyote, deer, otter, fisher, snowshoe hare, weasel, red squirrel, deer mice, vole, shrew, and ruffed grouse. A good day.

January 5
BOG TRIP

To get into Huckleberry Bog, in Chesterville, we walked down a small hill with a forest of maples, beech, and hemlock, before we came abruptly into the marshy flats of dry tan swale grass and speckled alders. Some of the alders held overwintering white patches of woolly aphids on their gray stems. Bushes of maleberry *(Lyonia ligustrina)*, with their tan winter twigs and tight bundles of round seed capsules, were a pleasing contrast to the grayish black winterberry bushes with deep purple twigs that still retained some crimson berries. One bush held the tattered remains of a yellow warbler nest; another, that of a goldfinch.

Walking a little further into the bog, we encountered the highbush blueberries with bright red buds and the telltale galls of a cynipid wasp *(Hemadas nubilipennis)*, that seem to emerge as smooth burs that bend the twig. The galls must, therefore, be formed as the new twig is growing. Among the blueberry bushes were low rhodora shrubs with the unopened pale lilac seed capsules of the current year, and the unfurling rich brown seed capsules of the year before.

These and other shrubs were interspersed with small, scraggly larch and black spruce trees. A few yards further, we finally cleared the brush and entered an open, flat area that seemed like yet another world. Here, on the floating part of the bog, the waterbed-like mat of interlocking shallow roots was covered with purplish green sphagnum moss. Over this moss ran a network of the creeping delicate cranberry vines, still bearing now delicious, frost-sweetened, juicy purple berries.

The ericaceous herbs grew only a few inches tall here. In almost any square yard, we could find the twigs with upright leaves of the bog rosemary *(Andromeda glaucophylla)*, whose thin, straplike, dark greenish purple upper leaf surfaces curled inward around snow-white underleaf surfaces. In the spring these plants would unfurl tiny pink bell-like flowers that attract bees. The bog laurel leaves had a similar shape to those of the rosemary, but their underleaf surfaces were not as white, and these plants would unfurl entirely different pink flowers in four months. Another member of the heath family, Labrador tea *(Ledum groenlandicum)*, had finely wrinkled green leaves pointing down along its upright stems, creating a tentlike space. These leaves, too, were curled around the edges, and their undersides were covered with a downy tan fuzz.

The retention of leaves through the winter seemed to be a conspicuous pattern among the bog plants, and no leaves were more noticeable than those of the pitcher plants. Their leaves are designed to catch and drown insects, by providing them a slippery slope of downward-pointing hairs into a deep pool of water. The students broke open the leaves, finding horn-shaped chunks of ice with insect parts trapped at the bottom.

The trees on this bog were bonsai-like black spruce, red maple, and occasional scrawny larch covered with lichens. Trees probably over 50 years old were only two or three feet tall, gradually growing higher toward the edge of the bog. The whole aspect of this bog was like a micro-encapsulated tundra, normally seen only many thousands of miles further north. Here we saw the latitudinal progression from deciduous forest to stunted trees to no trees, all compressed within a space of about a hundred yards. The bog also featured the same species of plants as the tundra, and it had been formed in a depression below the outlet of a small pond surrounded by tall oaks and pines. We next made our way to that pond on the ice of the connecting stream.

At one bend in the stream there was a section of thin ice, which we discovered after one of us broke through. We used the opportunity of drying socks and boots to socialize around a fire, and to crawl to the edge of our newly made "window" into the stream. The muddy bottom was only about a foot below the surface of the ice.

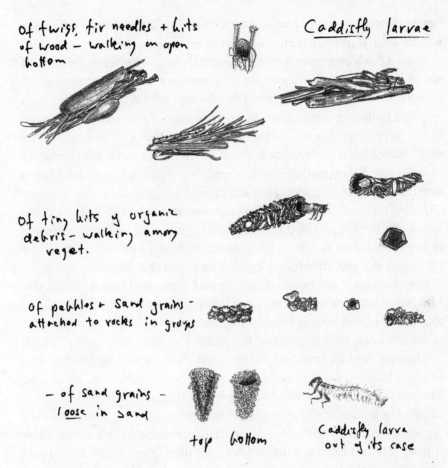

of twigs, fir needles + bits of wood — walking on open bottom

Caddisfly larvae

of tiny bits y organic debris — walking among veget.

of pebbles + sand grains — attached to rocks in groups

— of sand grains — loose in sand

top   bottom

Caddisfly larva out of its case

The view while lying face down, shielding our eyes from the sun with our hands, was superb. Whereas we had seen no animals at all so far, we now saw many small creatures, although almost no more plants. As we peered down through the crystal clear water onto the streambed strewn with bits of decaying vegetation, we saw a small movement here, then another one there. Caddisfly larvae, carrying their tubular houses of glued-together plant debris on their backs, crawled over the muddy bottom with their front ends protruding. (I saw two kinds, one with their houses made of large flat pieces glued together lengthwise, and another made of small thin pieces stacked vertically in a spiral.) When I reached down, they tucked themselves into the safety of their houses, and when I pulled one out of the water to place onto the ice, it quickly left its house. However, when placed into a jar alongside the house it had just abandoned, it quickly reentered its house.

I saw a back swimmer with the silvery sheen on its belly of a layer of air, its plastron, which is a gill used to extract oxygen out of the water. A rapidly swimming black gyrinid beetle zipped past erratically, then popped to the surface in front of my face, making a quick swirl before again submerging and disappearing under the ice. This probably would have been its only chance to skim the water surface for some six months. We fished out one of the beetle's predatory larvae, and saw a newt slowly crawling along the streambed.

A raven called as we ascended the stream toward its source at the pond. On the pond, dead pickerel were lying about near frozen-over holes bored into the ice. Ice fishermen had been here, and they obviously did not care for pickerel. The raven did, however.

For our lunch we had brought tinned tuna, and one of these still unopened tuna cans became a hockey puck. We plucked sticks from the top of a fresh beaver lodge along the edge of the pond, made goal posts from our backpacks, and the game was on.

After we had all returned to the cabin in the evening, Francie and Chelsea baked a fresh cake filled with our bog cranberries. Maria and Kim made bread, and we discussed a potential research project on caching behavior of red squirrels with Scott and Jessica as we warmed ourselves around the stove. Others read or studied their notes. After supper Jeff and Brad got up and washed the dishes, while David and Larry strummed their guitars upstairs in the dark. I loved the simple harmonic melodies, and wondered why I seldom hear anything as good on the radio.

January 6
A DISCUSSION

We are getting almost a minute more daylight per day now, but there is already a hint of spring in the air. This morning at dawn two male chickadees were calling their "dee-dah" song back and forth. In town I heard a good rendition of the song of a bluebird, but it was a starling on full throttle, mimicking a bluebird.

I went to the butcher in Livermore Falls today, as on most Wednesdays. Thirteen cows were accidentally electrocuted in their stalls in Avon, and they had been hauled to the butcher. The meat was condemned because the animals had not been slaughtered "properly" (whatever that is) and it would all be thrown away and others eaten instead. The butcher was keeping only the hides, so I could take all the meat I wanted. Using the blade lift of his tractor, Castonguay loaded a freshly skinned cow onto the back of my pickup truck. She barely fit, with some of her hanging out the back end. My bumper sticker—"This car stops for roadkills"—is more appropriate than ever. For the time being I park next to the students' cars down by the main road. Their bumper stickers say "Save the Rainforest," "Save Tibet," and "Practice Random Acts of Kindness and Senseless Acts of Beauty." Larry wears a sweater with "Earth First!" and a green fist on it. Earth First has just been in the local headlines—unfavorably—because of spikes that have been found in some trees slated for logging, and wearing an Earth First T-shirt in these parts is like walking around naked at Wal-Mart. However, I suspect hauling around bloody cow carcasses may be suspect as well.

EXCURSIONS

Jeff had previously sampled for water insects in a swift part of the stream where the ice had been thin. Now he needed a sample from slow water and we took turns with the ax chopping a hole through two feet of ice over my old swimming hole. The ice reached nearly to the bottom, where we swirled and scraped our insect net, and brought up netfuls of gravel. I had peered onto the bottom of the stream before, but I was not prepared for what was revealed in two or three scoops of the net. There were squat dragonfly larvae, camouflaged with algae and debris, but now fully revealed as they tried to crawl on the ice. There were squirming, naked mayfly nymphs that had always been hidden before. Now they showed their exquisite form, which has persisted almost unchanged for some 250 million years and was still near the ancestral insect form from which all others evolved. Mayfly larvae used and still use their gills not only to extract oxygen from the water but also to propel themselves through it by paddling. Ancient gills like these were the prototypes of modern insect wings.

An adult caddisfly

There were caddisfly larvae of three different kinds, and all were different from the ones Jeff had previously fished out from swift water nearby, as well as those from the bog stream. The exquisitely constructed house of one caddisfly larva was particularly intriguing. We would never have seen it by just looking at the bottom, no matter how close. While most caddisflies make tubular houses by assembling bits of plant debris, these were almost flat, consisting of a sheet of sand pebbles glued together to look just like the loose layer of pebbly gravel on which they were found. Under this carpet of camouflage they, too, lived in a little tube of silk.

After supper I asked the group if anyone was interested in a night woods walk. All were. The moon was hidden under a veil of clouds, and there was not a breath of wind. We walked in single file along the ridge of Adams Hill, then down into the firs where it was pitch-dark. When we came up and out of these woods to the edge of the hardwoods we stopped, stood quietly, and listened for a long time, hoping to hear an owl—anything. The woods were in absolute silence. For ten minutes, we didn't even hear the rustle of a leaf. Then we walked on. But before we left, Jeff gave an amazingly loud and accurate rendition of the barred owl. Again and again he called. There was no answer. I had not heard these owls since fall—strange, I thought, because they nest in late winter and early spring.

On January 11 we took Route 4 along the Sandy River, past the village of Madrid, past Rangeley and Oquossoc, over the Kennebago River, around Cupsuptic Lake, then around the south end of Aziscohos Lake by the Aziscohos Mountains. Finally we came to the Magalloway River, which drains the Magalloway Mountains into Umbagog Lake. Near where the Magalloway enters the lake, the Androscoggin River leaves it from an area of vast marshes. Here at the Maine-New Hampshire border we spent the day.

The Androscoggin still had open water in the center, but near shore the ice was at least a foot thick. The river ice was crisscrossed with cracks, and we heard occasional booming as new cracks were made. The ice was smooth but full of tiny air bubbles. There were a few small patches where a thin layer of windblown snow had collected and on one of these we found fresh coyote tracks.

As we walked up the river and got closer to Umbagog Lake, we entered a bog. At first we saw small tufts of grass, sedges, and leatherleaf sticking up out of the ice. A little further ahead there were also patches of sphagnum moss, then bog laurel, and more sedges. The carpet of sphagnum was colored in a pleasing mixture of pastel browns, pinks, and yellows. Waist-high black spruce and larch grew here, and the mat was interwoven with laurel and leatherleaf. There were many brown seed capsules of the blue flag iris on erect dry stems, as evidence of an even more colorful summer panorama. In my mind I could almost hear the snipe whirling overhead on warm spring days in this, their prime habitat.

Further off in the distance of the great bog there were trees. We saw a great white pine with a huge stick nest on top, probably that of an osprey. Using this nest as our goal and beacon, we made our way through the ever denser tangle of winterberry and red maple scrub. Finally we stopped under the nest for lunch, then moved on in the same line until we had crossed the bog onto the lake. Here on another great pine we saw the nest of a bald eagle. Nearby on a red maple, a red squirrel seemed oblivious to us, as it munched on buds just a few feet away.

The State Wildlife people had put sheet metal around the bottom of the great pine trunk to protect it. (The nest is technically in New Hampshire, making these New Hampshire's only pair of nesting bald eagles, but the next year the eagles renested a mile away, in Maine.) The Maine eagles aren't doing so well, but it's not raccoons, or greedy chick- or egg-snatchers that are the main problem. You know right off that when specific individuals are protected, there must be something terribly wrong with the ecology of their species. So it is here. Eagles along Maine's inland lakes, rivers, and marine habitats reveal some of the highest levels of mercury and PCB contamination ever recorded, and they are among the slowest to reproduce of any birds in the whole of North America. It's the chemicals: they affect pair-bonding behavior, and sexual development in embryos. A multidisciplinary group of experts that met in Racine, Wisconsin (July 26–28, 1991) concluded that "unless the environmental load of synthetic hormone disruptors is abated and controlled, large-scale dysfunction at the (human) population level is possible." A canary the bald eagle is not, but it should serve to warn us.

Someone noticed a minnow swimming close to the surface of the lake ice beneath our feet, and we saw mammal scats of several vintages all in one pile by a giant red maple leaning over the water. They could have been an otter's, but I wasn't sure.

We explored around the lake and returned by walking a raised bank or berm along the river, with the bog on the other side. The bog itself must have been part of the lake at one time, gradually becoming overgrown with the mat of vegetation as the river threw up the berm that cut off the flow of water. The berm was now covered with a strip of forest, primarily of balsam fir and red maples. It served as an animal highway, specifically for moose and deer. Every ten yards or so, there was another fresh pile of moose dung, and many of the branches that had snapped off along the path held moose hairs. We came upon a pool of urine that was not yet frozen, and about half a mile further we saw a moose walking in the forest ahead of us. I swung around it, trying to head it off so the students could see it also. The placid animal allowed itself to be herded like a cow, and I paraded it in front of an appreciative audience. Most of them had never seen a moose before, so it was quite a treat to get to within 20 feet of one in its native habitat.

The moose finally came to a stop where it was surrounded by ice, and as it took another step or two to get away from us, it slipped and fell. It slowly got up and tried to take another few steps, and again came crashing down, now to just lie there. We immediately walked away, leaving the animal to extricate itself in peace.

Leaving the raised wooded berm along the river, we reentered the bog to reverse our path back to the bay along the river. This time, as we entered the area that was half bay and half bog, someone found a spider walking on the ice. What might a spider now be doing here? Temperatures had, for several days, been near 0°F at night, and it was well below freezing now.

The spider opened our eyes. Suddenly we all found dozens—hundreds—of spiders, little ones and big ones. There also were beetles crawling on the ice in slow motion. I recognized one species that I had seen commonly on lily pads in the summer. But there were many other insect species, too, including a water strider walking on the ice. We all got on our hands and knees, watching snow fleas hop and the caterpillar of a noctuid moth trying to crawl. Might it contain sweet glycerol or other antifreezes? It tasted only marginally sweet—almost bland in comparison to the candy-like taste of overwintering carpenter ants.

As we traveled back onto the river, we saw a pair of gray jays. They were foraging along the river's edge, hopping in the low branches of red maples, then gliding with short, broad, outstretched wings in their curious, soft, owl-like flight.

January 13

What a thrill to wake up this morning to a deep gray sky, low temperatures, and tiny isolated snowflakes coming down! I knew from these signs that it would soon be snowing hard, and that the snow this time would continue and not turn to rain. And so it was. By noon the ground was already covered with an inch or so. The conifers were turning white, and the air, too, was now white from the increasing cascade of falling snow.

At this point in the seminar, teams of students had settled into doing their independent projects, after spending a week trying a number of alternatives. As a community project, I had them trapping small mammals to get an idea of the world under the snow. They caught red-backed voles, deer mice, white-footed mice, pygmy shrews, and Brad even caught a woodland vole and one smoky shrew. I had offered the Raven Feather Award—a long primary feather—to the first trapper to get six. Jessica, a petite blonde, got the prize. Several individuals from last year's class had refused to trap out of "reverence for life." Anticipating that reaction, I had told this year's students how we'd *facilitate* the natural process of recycling rodents into ravens. Since all loved the ravens, no one had any reverence-for-life qualms in this class, and all became eager trappers. We got a nice batch of deer mice and voles, at least 60 (or about a month of raven life). Brad and David even went on to make, as their project, a comparative survey of the small mammal fauna in a three-year clearcut versus a neighboring mature uncut hardwood forest. So far they had found that the clearcut was full of rodents, whereas there were significantly fewer in the mature forest. I encouraged the team to get a statistically meaningful sample. Every day the catch was put into the deep freeze—a plastic bag hanging outside next to the back door.

As each of the teams went their respective ways on their projects each morning, I made a quick run down the path to pick up the mail. I found a tall dark-haired woman with a braided ponytail and light green eyes wandering around down there, shouldering a case of Anchor Steam beer and looking a little bewildered. "Follow me!" I instructed, and she did ("It seemed like the sensible thing to do," she told me later), while I took the beer off her hands. She was a prospective grad student from West Hollywood, California, who had just flown the red-eye from L.A. to Boston, and had driven up here in a rental car that same morning. She wanted to see the ravens and the raven study site, and the professors and students at the University of Vermont, for possible doctoral work with me. This professor, at least, and the students at the cabin were happy to see her.

For our evening meals, people had spontaneously been volunteering to cook, with the aim to please. We'd had homemade pizza, macaroni and cheese, stirfries and rice. Every night somebody eagerly offered to cook something new and original. Now Brad was doing a batch of pressure-cooked baked beans—black, yellow-eyed, and pinto. Dave and Francie were making the bread. And today it was my turn to make a contribution.

As I said, we had a lot of voles in the deep freeze—more than enough for the ravens, who were at this time already feasting on fresh veal (stillbirths). I figured we had a few to spare. To prepare mice, you have to first thaw them out and pull the skins off and the guts out. Then you wash them, bread them carefully, and braise them lightly in olive oil in the black skillet. "Are you really going to eat these?" Dave wanted to know.

I had prepared a batch of about 30. By the time they were brown and crisp, his question no longer was quite so rhetorical. Clearly recognizing that life is art, Jessica nonchalantly picked one up and chewed it, neither flaunting nor hiding the fact. "Hey, these are *good!*" she pronounced. Jeff tried one and didn't comment, then pointedly grabbed for another, dipping this one in barbecue sauce.

Suddenly there was a run on, and mice were going faster than nachos and salsa. The second batch had barely begun to sizzle before the first was already gone. The Anchor Steam was going, too. Unfortunately, the prospective student decided that the University of Vermont was not for her.

January 20
COYOTES

The students are gone now, the thermometer hangs steadily between 0°F and 10°F, and I'm now tempted to yield to the urge to lie in the warm bed in the morning, then hover around the stove. But even though it is still dark, I hear a combustion engine in the woods across the valley. Some logger has been up for an hour or two, eaten breakfast, driven out under the starry skies, and started his skidder to warm it up. I get out of bed and start the fire, and after breakfast drive out to check on the ravens at the cows I had offered them.

A man beside a pickup truck was standing alongside the road, twirling a radio antenna. I stopped and backed up. "What ya trackin'?"

"Not ravens!" was his reply. "We're chasin' coyotes. Got one of my dogs radio-collared." He talked while looking off into the distance, over toward Carthage, not looking at me at all. I suppose he was thinking of his hounds, behind the coyote over there.

"You can't tree a coyote like a coon," I remarked, "so how do you catch 'em?"

His gaze did not leave the hills. "Sometimes we head 'em off and shoot 'em. Other times the dogs catch 'em—I keep puttin' relays of fresh dogs on 'em—eventually he tucker out."

"What do you get for coyote pelts?"

"Basically nothin'—we just do it for the sport of it, my partner and me."

I thought of the beautiful sounds of the coyotes I've heard so many times, and I saw in my mind's eye a coyote driven to total exhaustion, then torn apart by a pack of hounds. I also thought of coyotes chasing deer until total exhaustion, then tearing them apart. There was no point to my being judgmental.

"Good luck," I said, and drove on.

January 25
SWEET AND SOUR GRUBS

Last summer, within minutes after I had cut down a live pine tree to make room for sugar maples, beetles started to arrive—walking all over the felled log, mating, and laying eggs. The beetles are long dead now, but their larvae live on. First came the tiny bark beetles, whose larvae have now left intricate galleries just under the bark. I make a rubbing to copy the pattern. Next came the *big* beetles—the long-horned beetles.

Just under the bark I now see huge chewed-out galleries filled with stringy wood shavings, the leavings of the young beetle larvae. The larvae themselves are no longer under the bark. Grown larger, they have bored directly into the solid wood, leaving smooth oval holes to the surface. I chop just above and below a hole, prying off slab after slab of wood, thus pursuing the grubs into the interior of the log.

From eight feet of log I recover 23 grubs. All of them have been facing into the interior of the log, jammed tightly into their burrows and surrounded with hoar frost. Their creamy white bodies are covered with spiky white ice crystals. Temperatures over the last two weeks have often been near −20°F, though today it is "only" −10°F. All of the larvae are pliable, however; none is frozen solid. They are apparently laced with antifreeze, because they taste glycerol sweet, just like the overwintering carpenter ants also found inside trees. But I now wonder, are they really alive?

I bring five of the remaining uneaten grubs back to the cabin, where they should revive at 75°F next to the stove. Green lacewings and spiders that I've picked out from under bark in the winter walk almost instantly once held in the hand under a warm breath or two. Not so the beetle grubs, even after I squeeze them, palpate them, and breathe on them. They show not the slightest sign of that one essential attribute we think animal life should have: movement. How different from the cluster flies, I think, that come crawling out of the woodwork whenever I build a fire in the stove.

I kept the beetle grubs next to my bed in a jar with moist pine wood shavings, and checked it every day: 1/26—"Dead"; 1/27—"Dead"; 1/28—"Dead." I now stopped checking and nearly forgot about them. But on February 3 when I went to throw them all out, they were crawling around in the jar in quite lively fashion. I then put them outside overnight, in temperatures of about 20°F, or some 40 degrees higher than they had previously experienced outdoors. The next morning all five grubs were *solidly* frozen, unlike when I had first found them. I expected also that they'd be dead, and they were. I tasted one; it was no longer sweet, just nutty. The rest of the dead larvae turned brown two days later. Having lost their antifreeze after nine days in the cozy cabin, they were no longer protected. I suspect that cluster flies are without it as well, but I don't care to verify by the taste test.

THE SPIKER

Today my second five-gallon water carboy was empty. The last time I had filled the pair of them was exactly two weeks ago, when the students left. The water is low in the well, and solidly frozen over. I used the long maple pole to again pound my way through the ice before lowering the bucket down with the rope. I hauled the carboys up on the plastic sled, dragging it over the still shallow snow.

I did not really think I used this much water, but it is still likely a pittance in comparison to what I would use if I had plumbing. I have some suggestions for helping to solve the water crisis felt in some places out west. For starters, have the well a long distance from the house.

There is no lack of water here for me. I simply don't like to haul more than I need. Despite my reduction of maintenance activities down to probably one and a half hours per day, I'm still constantly pressed for time. The day races by and when I lie down to go to sleep, I review in my mind what I've done, to see if I've done the things I'd set out to do. Usually I haven't.

I'm finding that the challenge of learning the secrets of ravens gives me the most satisfaction. It animates me every morning to be out of the cabin at a run already a half hour before dawn, with a cup of coffee and some cereal under my belt. Lately I've run down the hill and crawled on my belly into a blind of spruce twigs, waiting for ravens to arrive at two dead skinned cows. Sometimes recognizing specific individuals is a thrill—like the one I saw the other day that I had marked on our first roundup seven years ago, and another that came from the nest at

Graham's Farm. I'm beginning to see patterns in their daily movements and we've mapped home ranges of at least 1,000 square miles. Also, I now know that at least 300 birds can be recruited to a super bait within a few days, even though only 40 or 50 visitors are there at any one time. And I'm developing ideas about some mechanisms that might explain how it's possible. Getting insights makes me feel high, and sometimes forgetful.

The other day I had pulled out the two large spikes driven an inch or two into a spruce as steps for raven watching, putting a ladder there instead. I left the spikes on the front seat of my truck, and then forgot about them. Suddenly it flashed through my mind: What if the spikes were found by someone who thinks that *I* was the self-proclaimed "Raven Watcher," the tree-spiking eco-terrorist? The newspapers recently had been full of letters saying that those responsible ought to be drawn and quartered. Then *another* flash: Suppose that the helicopter I had seen flying over—just before the story broke about tree spiking in this area—suppose it had had a metal detector, and had picked up the spikes in my spruce from the air—right next to Mount Blue, where the tree spiking had been discovered?

Strangely, the forester at International Paper refused to talk with me, when I called to ask a routine question. The Boise Cascade people had stopped work on my land, and *their* forester had left a note for me to call in my social security number to him so they could pay me—even though there couldn't be anything to pay yet. That was on January 26. He'd left the note in the cab of my truck that was parked by the side of the road! I hadn't called him yet. I frantically searched the truck—and everywhere—I couldn't find the spikes! I searched everything *again*, two or three times, as if not trusting my senses. I thought of my ravens, who I'd seen check for food where they'd been rewarded, and not finding anything, come right back and check again. I had thought this behavior indicated a lack of intelligence. Now *I* was doing it.

I sat down to think. No doubt about it—the spikes were gone from my truck. Fact two: I *knew* I had put them on the seat in my truck. Fact three: I knew that, besides myself, *only* the forester had gone in there, because at other times the door was locked.

Those spikes showed evidence of having been inside a live tree. What better suspect than someone with an out-of-state license plate, parked so very near the scene of the crime? What would someone have thought, had he looked on the seat and seen those spikes there? Might he think it's the infamous "Raven Watcher"?

I told the whole bizarre scenario to Ron and Syndi. *They* thought it was slightly more humorous than I did, but I laughed, too. Ron nevertheless advised that I take some pictures of the shallow holes where I had pulled the spikes out, "Just in case they don't mysteriously get deepened, to make it look like nails had *really* been driven in." After all, I'd already gone on public record as questioning the concept of calling herbiciding "forestry." I still hadn't called the Boise Cascade forester (with whom I consulted about the cutting on *my* land), who I now thought had been the one who took the spikes. Why hadn't he *told* me? Should I *ask* him? No, I decided I'd play innocent. See how far they'd take this. Despite being innocent, I started to see myself a little bit as I felt *they* would see me if the gossip started flying, with hearsay and innuendo that nobody could prove or disprove. Ron said, "You've been out in the woods too long—that's what happens to people if they're alone too long. They get woods-queer." I wondered.

In the afternoon I needed to get out, just to talk to somebody else. I went to see my mother. Marianne was there, visiting. I gave them the details.

Marianne concluded, "Brother, you're *had*. If I were you I'd get a lawyer right away. If they're looking for 'the spiker' you've got all the credentials."

Mother was more positive. "We'll come see you in jail." She smiled. "Henry Hilton (our mutual friend and neighbor) went to the Farmington jail for a month. They took him because he refused to pay his wife—after she took off and left him with their child and moved to Arizona. He said he *loved* it in jail. They fed him. He didn't have to work. He spent all his time drawing pictures. *You* could write."

On reflection, going to the county jail might be relaxing. Thoreau had done it, for a short stint. I had too, on a Fourth of July weekend when four of us from the woods crew in northern Maine had hitch-hiked to Quebec City. We didn't have enough money to buy lodging, it was raining hard after midnight, and our sleeping bags were soaked. The drunks' moaning kept us awake, but they let us out the next morning at 7 AM, on demand, unlike eco-terrorists. Eco-terrorists, they would keep.

January 31

After the last couple of nights when the wind howled and shook the cabin, last night it was deathly still. In half sleep at 10:30 PM I found myself changing position from a sprawl to holding my legs together, to crossing my arms against my chest, to spreading my hands against my body and drawing up my legs. As it was getting colder, my body was automatically adjusting to reduce its surface area for heat loss. Eventually I got up and put more wood in the stove, noting an almost eerie blue moonlight outside. Next to the window I felt the sub-zero cold fingers of the arctic chill stabbing into the cabin, and I also found them reaching in between the logs.

At 12:25 AM I awoke again. This time the night's stillness was broken by the mournful deep howling of a coyote, or were they wolves? The long slow howls would have fooled anyone. Timber wolf is what they sounded like, their deep and resonating power now trailing off into musical tremolos. Far in the distance there was an answering call. Then silence again fell like a curtain. I put more wood on the fire and slept.

The mice came back. One recently had been putting on a strange display. It made short runs, and at each stop I heard a very brief buzz that sounded like some giant fly. It was a vibration probably made by a rapid stamping of the feet. I had never heard this before but this mouse was doing it nightly for the hour or so that it was active long after midnight. The behavior had, as far as I knew, not been described before. Was it a signal—like jungle drums—to try to contact another mouse?

In the very early morning I was again awakened, this time by the ripping sound of a small avalanche of snow sliding off the roof, just a few feet above my head. This was followed by the barely perceptible rustle of snow crystals hitting the newly exposed portion of the roof.

At 5:30 AM the jagged jangle of the alarm clock rouses me, and I retreat by crawling deeper into bed. But then I think of the ravens: I must follow through with my observations. And I am revived by the thought that it will all be bearable, if I only have a hot cup of coffee. I now leap out of bed, and in the cabin soon illuminated by the faint yellow glow of the kerosene lamp, I start a roaring fire. The temperature inside the cabin is 20°F, and I have to throw out a solid block of ice from my porcelain pan (wash basin), but in half an hour I have my coffee. A miracle.

On my way to the outhouse I read the outside thermometer. It reads −20°F—40 degrees colder than inside—a consolation. And for a very few minutes I listen to the whisper of tiny tinkling snow crystals falling now in ever denser sheets. Winter has *finally* arrived. Until now we've had low temperatures and we've had precipitation, but the two have usually alternated. Today, for the first time this season, they promise to combine.

By the time I'm trotting down the path in the pre-dawn darkness, feeling the snow on my face and hearing the squeaky grind of the very cold snow like the texture of sand under my feet, I'm in a great mood. I'm off to see the ravens from my spruce blind.

It is so cold this morning that even some of the ravens, while walking in the snow, on occasion pull first one foot and then the other up into their fluffed-out belly feathers. All have white-grizzled faces, where the moisture from their exhaled breaths has frozen onto their feathers. The constantly falling snow itself is too dry, and its crystals are thus too small, to stick to their backs.

As always, the ravens come near dawn. One or two of them yell for half an hour, and then 20 descend to feed. After that, as I hunker down in the dark cave of snow and branches, I hear only the pounding of their bills on the rock-hard meat of the remains of the cow carcasses. Only two marked birds have come today. After 90 minutes my feet are numb and I'm shivering violently. I'm forced to withdraw from their feast, to return for a warm-up at the cabin.

Got a letter from a friend yesterday who wants me to accompany her on a tour from Anchorage to Khabarovsk, to then travel by the Trans-Siberian Railway to Lake Baikal, "home of 1,700 species of which 1,080 are found nowhere else," all for about $2,000. I feel no need, nor great desire. There are many species here, and I hardly know them. I get sensory overload from the couple of dozen species of mosses alone. One species, the ravens, already blows my mind.

February 2

It's Groundhog Day. The groundhog is known here as the "wood-chuck." Woodchucks, like many other ground squirrels, hibernate in their underground burrows where they are thought to sleep away the winter. Actually, it's not quite that simple. A friend and colleague at the University of Alaska, Brian Barnes, has found out in his extensive studies on the hibernation of arctic ground squirrels that these animals stir from their cold-bodied torpor at numerous times throughout the winter, *in order to sleep.* Given the long nights, I too sometimes feel I must wake to warm up, just so I can go back and sleep soundly. Sleep is associated with certain brain waves that only occur in a warmed animal. Apparently the animal can go all winter without food, but it periodically needs real sleep.

According to the German tradition brought to the Pennsylvania hills in 1871, if the groundhog sees its shadow on February 2 it will be frightened and return to its hole for another six weeks. However, so far in 107 years of prognostication in Punxsutawney, Pennsylvania, the very experienced "Punxsutawney Phil" has *not* seen his shadow, thus predicting a thaw in two weeks, on only ten occasions. The first time was in 1890 and the last time in 1950. The groundhog's accuracy is further questioned by Accu-Weather, Inc., a commercial forecasting service. *They* say February and March are about as likely to be warmer than normal as they are to be colder than normal which sounds suspiciously like the law of averages.

Well, no groundhog will see the top of the ground today, nor any shadow if it did. Not here in Maine. It snowed without a break all day yesterday, all through the night, and all day today, and temperatures are remaining near 0°F. Writing for the *Lewiston Sun*, Karen Kreworuka says that according to her dear friend and neighbor, Vernon Hutchinson, Groundhog Day falls about halfway through the winter, and you can judge the prospects for survival on three criteria: you still should have half of your firewood, hay, and root crop left. I've got slightly less than half of my wood used up, but I'm burning it up ever faster.

If the weather is any indication, this is not halfway through the winter. This is just the third day. This is the first day I've walked in deep powdery snow up to my knees, seeing the fir and spruce branches bending under more and more falling flakes that, because of the low temperatures, promise to stay around for a while. The woods look like a fairyland, but only from the perspective of one who has a warm fire and an ample pile of dried firewood.

I was helping Syndi shovel out the mailbox and parking spaces down by the road. A pickup truck with a plow rolled by, driven by a long-haired man and his little boy. A hundred yards down the road they stopped, backed up, and without a word started plowing our drifts away.

Syndi turned to me. "Do *you* know this guy?"

"No, I've never seen him in my life."

There's a sign at the Bangor airport: "Welcome to Maine—the way life ought to be." That's just the way it is.

Called Si Balch, the forester, today to give him my social security number, as asked for. He was cool. He didn't say anything except that the cutting had been delayed. Dale, his logger, had cut two truckloads of wood for me, and that was it for now.

I thought back to Marianne's response when I had told her the story.

"Let me get this straight," she'd said, "it was early in the morning—before daylight—that the forester just by chance happened to come by and see your parked truck near the scene of the tree-spiking crime—and he opened your cab to drop you a note, and he saw two spikes about a foot long, that showed signs of having been in a tree—and he *took* them, and he told you to call him to leave your social security number. That's all?"

"That's about it."

"Brother, they'll *get* you."

COWS

The powdery snow flies up as I trot through it, determined to make it to the blind before the ravens arrive at dawn. The blind is now covered with deep snow, making it a fir-lined igloo! But before I duck in, I pause a moment to gaze at the rapidly changing color of the sky. It is light turquoise in the east, grading to deep, dark, brilliant blue in the west. The colors fairly shine through the black silhouettes of the bare aspens and the thick branches of the firs and spruces, laden down with their gleaming white burdens.

Coming back up the hill later in sunshine, my eye catches the glint of millions of pinpoints of reflected light from the snow crystals. Each little beam emanates from a facet or tiny mirror of a flake, and vanishes from view at the slightest change of angle between it and my eye. But while some lights inevitably turn off with each step I take, just as many others turn on, making the whole surface of the snow twinkle constantly as hundreds of thousands of mirrors flash on and off at once.

Here and there a red squirrel has tried to travel in the fluffy snow, but has managed merely to tunnel along. I understand at a glance now why the squirrels cache their apples, and why they do this far up in the trees rather than on the ground.

*   *   *

I treated myself to "the usual" at the diner in the afternoon, as a reward for having spent most of the morning going over a draft of a technical paper on ravens for an ornithological journal. The diner was not very busy, and sitting at my usual place on a bar stool close to the cash register, I chatted with Roger, the dishwasher.

In the kitchen through the window, I could see Mike, the owner, busy with the cook and two other helpers. He was jabbering away in Greek. Then he came out of the kitchen, taking a break himself. He always wears a black visor cap with some small lettering on it. He has thick black eyebrows, dark brown eyes, whitening sideburns and a pencil-thin mustache. He was talking with a customer next to me, telling him how unusual the weather had been—not enough snow— and how few skiers had been stopping by (on their way to Sugarloaf).

A customer was having a long conversation with Lauren, the waitress, who was just starting her shift. Lauren had been having problems. Her car was broken down, and her landlady wanted it off the premises *now*—or Lauren and her kid would have to *go*. Lauren was upset, but the customer advised her to talk to one of the sheriffs when he came in for his coffee, to find out what she could do. The customer said Lauren had a right to park her car on the premises, and that if the wicked landlady objected, to stop paying rent, put the checks in escrow at the bank, and give her copies only. "She *can't* just throw you out," he advised. The other waitress, Chris, had had the same problem last year with *her* landlady. I decided that Farmington was not a good place for renting.

The blond cowgirl and her cowgirl friend were at the opposite end of the bar, and after my free dessert (a feature on Tuesdays), I walked over and asked her if she had any dead critters I could have for my ravens. She didn't have any at the moment, but she gave me the name and location of a farm where I might find something.

It was another nine miles past the diner to the farm. The road was hilly and winding. There were patches of snow on the road, and a few flakes were still coming down. Thickly clad snowmobilers with goggled black helmets whizzed by along the banks on the side of the road.

The barn where the Holsteins were kept was not like the ones where I'd milked cows, shoveled manure, and pitched hay when I was a kid. This was a new barn. No handhewn timbers and steamed-up, iced-up windows. In fact, there were no windows at all, nor even any sides. Just a cement floor and a metal roof held up by a wooden frame. The cows were in two inward-facing rows with hay and silage spread out in front of the steel bars where they were hitched. The dead cow lay in the center, just in front of them, between the two rows of live ones. She looked kind of bloated, but the cows chewed on unconcernedly.

The ruddy-faced farmboy with a slight stubble and a visor cap on backward seemed eager to drag out the cow with the tractor. So we hitched my chain around her neck and onto the tractor bucket, and he hydraulically lifted all 1,300-odd pounds of her, drove to my truck, and began to lower her into the bed. I noticed now that her undersides were a vile purple and red, and then I got a whiff that made me double over and gag, scrambling for air. I expressed doubt that the ravens would eat her.

"They probably don't eat it *until* it's rotten," he speculated.

"No, actually they like it fresher."

"Well, this one is only two days."

That must be biblical days, I thought.

"Can't we just dump her in the woods near a road here somewhere?" I wondered. "Then I'll come and see what ravens end up here, after I cut her up good."

"We dumped one once, but the landowner didn't like it. Where do *you* dump 'em?"

"Alongside the road in the woods."

"Yours?"

"No—but nobody has ever complained. This one, though—I think I'd better pass her up. Looks like she won't fit in my truck anyways."

"OK—how about some calves instead?"

The six young calves in a pile were all frozen solid. We loaded them on.

"Another feller came the other day to get some, too. He uses them for coyote bait. He nooses them. Caught 20 in the same area. He says there are even more there this year."

The Maine coyote is on the most-wanted list. In the spring of 1989, the Maine Legislature, realizing that bounty hunting has a bad connotation, established a Coyote Award Program, to "control" the animal. First prize was an award of $1,500 for the biggest female killed during the year. Other awards followed: second prize ($1,000) for the largest male; $500 for the highest number of males or females, respectively; and another $1,000 for the *total* greatest number, regardless of sex, that anyone might "control." The award program was scrapped, but there is still a year-round hunting season, and even special night-hunting periods. And all this was enacted even though state biologists had advised against the measures. They had concluded that, given the animals' biology and this state's terrain, the coyote is here to stay. They had shown that the greater the number killed, the bigger the average litter size would become.

The farm boy wanted to know if I was one of these coyote hunters. I admitted that I had no "useful" purpose in mind such as controlling coyotes, and that I simply wanted to feed ravens to watch them.

"You do this on your own?"

"Yes, mostly."

Coming back through Farmington, I stopped at the diner again. Joey was just starting her shift now, and she was singing while making her way around the tables.

"What will you have, honey?"

"Oyster stew."

EARLY FEBRUARY

The Black Bears, the University of Maine hockey team, last night extended their unbeaten winning streak to 41 games. They are now ranked number one in the country. The paper says, "As the brutal New England winter drags on, one of the few warm spots is on the ice where the top-ranked Maine Bears dominate their opponents. The Bears have become a source of pride and exhilaration for a state that doesn't have much to boast or cheer about." I wonder where they get their information.

It's near −20°F again, and clear and bright. The full moon last night was so bright I could almost read by it. I again heard the call of a barred owl—the first one I've heard in months. Like the great horned owl, these owls should now be getting ready to nest. They lay their eggs in midwinter, incubating their eggs and chicks through many blizzards.

Today, also for the first time in months, I again hear blue jays screaming. Not since fall have I heard such a ruckus. A pair of ravens circles high overhead, making a few soft grunting sounds. One of the birds repeatedly dives and twists like a corkscrew on its way down. They, too, are getting close to the time when they start to build their nests. In two weeks some will have already built them. Spring is in the air.

And in the spirit of renewal—I found the two missing spikes. They were in a grocery bag up at the cabin the whole time. I remembered leaving them in the truck after pulling them out of the spruce when I set up the ladder. What I didn't remember was picking them up later with a load of groceries, and then leaving them in the empty bag. So much for the mystery, and my paranoia, and my chance to defend myself in court against rumor and innuendo.

March 13
STORM WARNINGS

It is a beautiful, clear, sunny day. But the paper headlines the forecast of a "monster" storm for tomorrow. On my way to the diner, I switch on the radio—they really are talking it up. And at the diner, Mike confirms it. "Storm of the century," it's called. Maybe I'll get socked in for a while. Better stock up.

At the Shop 'n Save it looks like a fair. All nine checkout counters are backed up with huge lines of people pushing loaded carts. You'd think they were laying up supplies for the invasion. There is a jolly atmosphere, as though *finally* they'll see some *excitement.* I get a bag of potatoes, some butter, and a few more cans of red beans.

In the night it was deathly silent. I could hardly remember it ever having been *so* still. When I awoke at 6 AM, the sky had turned hazy. It was still quiet.

At 10:30 AM I put on my snowshoes and went for a long walk in the woods. There was still not a breath of wind. The sky was now a light gray. The snow was powdery and deep under my feet, and it clung in large cushions to the branches, from a previous storm. The cold, deep silence and the crunching of my snowshoes as they sank eight inches deep with each step reminded me of my boyhood and going into the winter woods alone to look for animal tracks, my mind aflame from the stories of Jack London. I'd stop and build a small fire to warm myself. Today I did so only in my mind. I continued walking, seeing many deer tracks on the south-facing slope of the hill that was covered with half-grown spruce and fir. The tracks were not as bunched into paths or highways as I had anticipated. Wherever there was a sapling sticking above the snow, deer tracks led to it, and I could see the yellow snip marks of the wood, where the terminal part of the twig had been taken.

Birds were few. Only twice did I hear the cheeps of chickadees. But I flushed a barred owl from a red maple tree among the balsam firs. It flew as lightly as a feather, and as silently as a ghost.

By 11 AM I could just barely see tiny snowflakes against the black conifers. But hour by hour they came down more densely, and very gradually a wind picked up.

At 5 PM, the trees were swaying and heavy snow was being driven in sheets. I heard a low roar and a sound like sandpaper rustling as the snow was driven against the cabin. The forecasters had said, "Meteorologists refer to this type of explosive development as a meteorological bomb," and, "They don't get much bigger than this," and, "People better pay attention—we're not fooling around." Up to three-foot accumulations, and drifts of 10 to 15 feet are expected, and winds should increase to over 60 miles an hour. I couldn't wait.

Weather experts and climatologists say we're seeing the aftereffects of the violent eruption of Mount Pinatubo in the Philippines in June 1991. This was the most powerful eruption the planet had seen since Krakatau, near Java, erupted in 1883 and created weather disturbances for more than ten years. Mount Pinatubo erupted for 24 hours, and during that time released force equivalent to the detonation of one atom bomb every second.

March 14
STORM

A little after 5 PM last night the wind did pick up, and it got fiercer by the minute, driving a fine, granular snow. You know something is up when you're sitting on the couch next to a roaring fire, and you feel a cold breeze. The chinking! It wasn't tight. This had happened once already, and I thought I had found all the offending little cracks then. Luckily I had a big case of oakum downstairs. Normally it is used for stuffing cracks in boats; I now feel like I'm plugging an old leaky boat myself. The job has not been done to Admiral Hyman Rickover's specifications. He admonished, "Nature is not as forgiving as Christ."

As the wind got fiercer, I found nature increasingly unforgiving as the cold fingers of the storm continued to penetrate, reaching across the room in icy plumes of snow that accumulated in drifts on the floor. I didn't know if I was being brave or foolish. But I felt no fear, only exhilaration.

Near 7 PM, after an hour or so of steady chinking and continued rough weather, I called it quits. With my hands I could still feel cool air coming in along almost all the cracks. Well, at least most of the snow coming in was staunched. Yet now I wanted to feel it outside, to embrace the full force of the storm.

When I tried to leave the cabin, I found the door already blocked shut with snow, and I had a hard time pushing my way out. More difficult still was the task of putting on my snowshoes in the dark. Going by feel, I needed bare hands to find, fasten and tighten the leather bindings. Mostly what I felt was numbing cold, but I got them on just in time before sensation left my hands entirely. I put my hands back in dry mittens and feeling soon returned.

As I left, almost wallowing through the woods, the snow was falling almost horizontally, stinging my face like sharp little needles. So I turned my head down. The light of the moon and stars had been blocked out. Normally the snow looks light against the pitch-black trees, but today even the snow was almost black.

In the distance came a roaring like a giant train going down a tunnel, speeding up to a whistling momentum, then momentarily slowing down. However, this roar came from *all* directions. The distant roar was the wind playing on the trees.

As the pitch rose and fell, the trees themselves chimed in with groaning and creaking. Limbs and trunks rubbed together and intermingled, humming deep tunes like giant bass fiddles. Percussion sounds came from the constant clicking and clacking of the branches, knocked together when the trees hit one another, bent violently, and then quickly rebounded. Above these sounds, from up close, you heard the fine, thin hissss . . . of the snow driven against the twigs.

As I walked along with my head half down to shield my face from the wind, I suddenly noticed two large black forms bound past, not more than 15 feet in front of me. Deer, probably. I could not make out any features. Within a second or two, a third one came by, following the first pair. This one almost ran me over. Definitely a deer.

The roaring continued well into the night. But when I awoke in the morning, the storm had passed. I dug the calves out of the deep snow so the ravens could feed. A bird flew by and saw one of the half-eaten calves that I had dug up. It yelled, and within five minutes, I saw four other ravens hanging in the air above me.

No hint of the sun all day. Just gray skies, continuing strong winds, and swirling deep snow. I'm exhausted from just walking. I do not find the kinglets today. I see almost no birds at all. I feel strangely depressed, with no energy to *do* anything. After I come back from the woods, I sit by the fire and stare, and drink a beer, and slide even deeper into a torpor. I dream of the returning spring birds. Spring? Will it come? This is almost too much. The snow is so deep that walking in the woods is an exhausting effort, and I'm like a bear in his lair, loathe or lazy to leave it. Motivation is also down, because I have finished the experiments with ravens, having answered the questions I set out for myself last fall. Until now the hard discipline to follow through has rousted me outdoors on sub-zero days even before daylight. Now I need some other focus to give meaning to every day. Life is not a spectator sport.

March 14
AFTER THE STORM

I got up but didn't even bother to make a cup of coffee. After getting dressed and putting on my snowshoes, I headed down the trail for a good breakfast at the diner and to see some *people*.

As I sat down on a stool at the counter, Joey immediately put a cup of coffee in front of me. A man next to me talked about his wartime experiences in Europe and told corny jokes. I laughed. I told equally corny jokes that I remembered from my days with the woods crew, up north above Mount Katahdin, in Aroostook County.

By the time I got back to the cabin, I felt revived. All afternoon I walked on the snow of the monster storm. Thanks to snowshoes, I could walk on water. Frozen water, that is.

Deer tracks were everywhere down by the cutting. At one place I saw six fresh, deep beds close together in the snow. So there were quite a few deer enjoying my new "bed & breakfast"—the small logging operation. Surprisingly they still were not yet keeping to trails—they were not "yarding." They went to wherever there were cut-down branches, nibbling off the buds.

There were other animals that had left tracks on top of the snow, not sinking in at all. I saw several tracks of rabbit, grouse, and one weasel. At a little clearing by the brook, a grouse had popped out of the snow, having spent the night in a little snow cave secure from the storm. But my main goal still was to find a miracle—a live golden-crowned kinglet. It was −13°F this morning. Could this bird, which when plucked is no bigger than the end of your thumb, keep warm enough to survive?

I searched for five hours. Once I *thought* I heard the birds' thin cheep. But I wasn't sure; their calls are subliminal, hard for me to distinguish sometimes even from the mere *thought* of a faint cheep. Suddenly I saw a small olive bird fly past, and I chased after it. I spotted it briefly as it hopped among the thick branches of a balsam fir close above me. Within a few seconds it flew on again, vanishing from sight and hearing.

March 16

Alder Brook is a fair-sized stream but you don't see any hint of it now. There is not even a gurgle. Temperatures have been near −10 to −20°F for months now every night, and the deep snow has obliterated even the banks. It is hard to believe that beneath my feet there are colorful fish and insects of all sorts. There are signs of life above the snow, too. Deep furrows containing much deeper foot-holes meander back and forth across the brook between the fir-aspen thickets on one side and the towering pines with hazelnut understory on the other. The furrows look like a herd of giants have been plowing hither and yon through the snow. It was moose. They were here yesterday. There were four of them; I found where they had bedded down as a group, within about 50 square yards.

In the spring, the brook is one continuous torrent and in the summer it becomes a wild cascade of riffles and swift water, interspersed with pools and shallow falls. The woods on the other side slope up toward the spruce and then the bare ledge summit of Mount Bald. This is a deciduous forest of red oaks, beeches, sugar maples, white birches, ash, and bigtooth aspens. Scattered through it are a few white pine trees well over five feet thick at the butt. As I continued through the woods, on my search for raven nests, the sun shone brightly onto the deep glistening snow. Although temperatures had been sub-zero early in the morning, the sun had caused considerable warming by early afternoon. I walked laboriously, sinking deeply into the heavy crusted snow with every step. First I checked the lone pines, by orienting them with respect to the sun. I found no large opaque areas that would indicate a raven nest.

In the valley that roughly parallels the road and the stream is a very long, sinuous, and narrow hill. Walking along the top, you look almost straight down on both sides. I picked my way along the narrow ridge, an esker, following a coyote who also had walked there recently, probably because the footing was too steep on either side. A few white pines and some hemlocks grow along the top and the sides of this esker. I again found no raven nest, but I saw a porcupine perched on a hemlock limb that it had already pruned of twigs. It looked at me without moving. When I passed by on my way back an hour later, it was still sitting in the same spot, looking at me again.

As I was about to leave the woods, I came upon a flock of several chickadees, a downy woodpecker, and two or three red-breasted nuthatches. One of the nuthatches perched upside down on a dead pine stump, hammering vigorously around a small hole. It put its bill in as far as it would go, but pulled nothing out.

Two ravens gave violent chase to a third. And then four of them flew together toward Mount Bald. Two came right back. A pair of trespassers had been escorted out, or two neighboring pairs had briefly met.

When I returned to the cabin at 3 PM, I was hungry, tired, and bathed in sweat. I stoked up the fire, put in two large potatoes to bake, and within minutes I was napping soundly on a soft, warm bed close to the fire.

SPRING

March 20–21
FIRST TAPPERS

The bright sun and the warm stove heating the cabin are bringing out more flies from the woodwork. At least five times per day for the last several days, I've opened the windows to let swarms of them out. You can tell how cold it is by how far they fly before they crash. I still don't willingly share my coffee with them in the morning, and I still haven't come to the point where I can cozy up to them in bed.

Last year I had seen, just like yesterday, a red squirrel behaving very curiously. It scampered across the snow, leaped onto one of the young sugar maple trees, and then began to lick under one branch after another. Sometimes, however, it didn't lick at all. Instead, it bit into a limb and instantly ran on. The squirrel did not chew bark to get at sap or something else. It merely made a quick snip.

Biting a tree and then running away seemed like a very curious and wasteful behavior. What was going on? I examined closely where the squirrel had bitten the branches, and found the bite marks in the thin bark. Two tiny slivers of bark were still attached where the two teeth had angled into the limb, so it was clear that nothing had been taken. I found these seemingly senseless bite marks by the hundreds. And when I made extensive surveys I found them almost exclusively on sugar maples.

Were they simply licking sap? No. Sugar maple *sap* is nearly tasteless and about 98 percent water. If the squirrels were to drink it to get the sugar, they'd soon get bloated.

I eventually concluded, however, that the red squirrels were harvesting maple syrup, much as we do. They use the same principle: wound the xylem to start the sap flow, wait for sap to collect, and then evaporate the water. But they have combined the last two steps into one.

Like a human tapper who drills a hole into the tree and then leaves to let the sap drip, a squirrel also leaves instantly after punching in a hole with its sharp incisors. However, it remembers where it made its tap, and by the time it gets back, possibly the next day or so, the sap has collected and evaporated. The sticky fluid adheres to the bark and runs down in a long streak. A small drop of sap, when spread out over a large surface area, evaporates rapidly when the weather is just right. And the weather *is* just right for evaporation on those days when the sap runs, namely when the nights are cold and the days are warm and sunny. (The cold temperatures at night ensure that the air is dry, and the warmth of the following day means that this same air can now take up much moisture.)

The squirrel remembers where it made its taps, waits for a warm and sunny day, and presto—it is eating maple syrup. I watched the squirrel run from one tree to the next, and it went straight to those—and only those—that had taps, where I then also scraped off thick squirrel-made maple syrup and candy.

At face value, this is "nothing but a cute natural-history observation," as a reviewer of my paper said when I worked this out and sent it to *Science*, where it was rejected. But it seems to me that the principle that red squirrels *make* food, essentially out of nothing, is *magic*. The magic here lies not only in the identification of the right tree, but also in the mechanisms for getting the sap and then making the syrup. (In my ignorance I tried to mimic the squirrels by wounding twigs with my jackknife, but I got nothing. It was only much later that I realized that *maple* sap only flows from injured *xylem*, which is the wood layer *beneath* the bark, in contrast to the sap that the yellow-bellied sapsucker gets in the summer from *other* trees.) Evaporating the maple sap, in the case of the squirrel, is a passive undertaking, if you call patience passive, which as a raven watcher I'm not yet willing to concede.

How did humans figure out that you could get sugar from a maple tree? I'm sure that, like most discoveries, it was not funded by a research application that spelled out to the last detail what was to be discovered. No amount of knowledge and inventiveness would have been sufficient for a human being alone in a forest to point to a tree and declare, "*This* is the tree I want to get delicious maple syrup from. And I'll get it by puncturing the xylem to collect tasteless water. I'll then evaporate the water off and pour the product on my pancakes and call it 'maple syrup.' " You find new things by rambling, not by racing.

Ron and I think of sugaring on a small scale, driving to Hall's Farm in East Dixfield to buy a few supplies. As we drive by the end of Wilson Lake on Pond Road, we see a progression of little houses being pulled off the ice. The ice must be softening, and the ice fishing is close to the end. I've not yet had a chance to use my new ice-fishing equipment, and now I'm already thinking about maple syrup. I buy 20 metal sap buckets with covers and spiles, for $2.50 a set.

The Native Americans collected the sap in containers made from the bark of white birch trees. They grooved the bark of the maple trees, nearly ringing the trees to get the sap out. Now we merely drill a $\frac{7}{16}$-inch hole and drive in a spile from which we hang a pail, although most modern maple sugar farms connect the trees by plastic tubing, and then pipe the sap directly into a saphouse where a suction may be applied to draw the sap out even faster, and where giant evaporators boil thousands of gallons of sap daily over super-hot furnaces. The Indians laboriously boiled the water off by repeatedly dropping heated rocks into the sap held in large white birch-bark containers.

There has been tremendous evolution in our technology of maple sugaring. But it all got started from the seed of an idea planted in someone's head. That seed grew because its fruit was sweet. But where did the seed originate some thousands of years ago? Could it have been planted by a red squirrel?

I suspect that, had I *not* known about maple syrup, I *might* have learned about it that first time I saw the squirrel's curious behavior. Lest you think I'm flattering myself, even chickadees catch on to the squirrels, flying from one branch to another where the busy tappers have done their work.

Sugaring time is when the nights are still cold and the days are sparkling and warm. That time is now. The sky toward the north on clear days is such a deep, gorgeous blue that I have to stop and stare, as if I have never seen anything like it. This blue is not the proverbial robin's-egg blue; this is more like the hermit thrush's eggs, or even those of the catbird.

Perhaps the color of the sky is no different in the summer. But in the summer, your eyes confront a riot of color and hardly notice another bright one. For months now, there have only been grays and browns; dark, almost black greens; and an ocean of blinding white. Under the sun, the white of the snow is so dazzling that you actually look up into the sky to rest your eyes on something darker, for relief from the brilliance.

In the upper layers of the snow, the sun's warmth is eating away, melting and coalescing the flakes so that they freeze into a granular crust at night. Soon the crust will be hard enough to walk on early in the morning—then you know spring is truly on the way. Since the day of the big storm, the wind swirling around the trees has blown the snow out in circles all around every trunk. This lack of snow close to the tree stems looks like "melt-holes"—an easy thought on this sunny day.

The beginning of sugaring is a much surer sign of spring than is Punxsutawney Phil. But the real sign of spring is frogs. And in my half-sleep, I drowsed off dreaming of the croaking of the wood frogs and the piping of the spring peepers. The frogs' voices rang as clear and precise, to my mind's ear, as if I had been at the pond. Right now, however, these frogs are not yet here. Instead they are crouched with their legs tucked in under their chins, hidden under damp leaves all over the forest, and all still covered by three or four feet of snow. They will not move a muscle for at least another month or two.

March 23

Yesterday afternoon I drilled the 20 holes, sunk my 20 taps, and hung the 20 buckets. No sooner had I pulled out the drill of each hole than clear sap started to ooze. Pounding in the spile, the sap dripped almost immediately. By evening there were about two inches of sap in the bottom of each bucket, and I went around with my plastic toboggan and emptied the buckets into a carboy on it. Then I fired up my rusty barrel stove, to test out the evaporator. Within minutes it was billowing steam. None of this was remarkable; that's the way it was all *supposed* to work. But somehow it seemed like a miracle. There had been a hundred things to attend to in order to come to this point. I had been pruning out the maple grove for several years now. The sugar shack also had taken some planning. Last summer I had hauled red maple trunks out of the nearby woods for the frame, and dragged up boards purchased from Parker Kinney's lumberyard for the walls. I'd boarded up the shack, with Stuart helping me and having a great time pounding nails. I'd covered it all in tarpaper. I'd cut firewood with the chain saw while the blackflies were out; then I'd dragged and split and piled it. There was the barrel stove that Ron had picked up at an auction, as well as the evaporator, the spiles, the drill, and the buckets that I'd purchased and had to lug up. Finally, I had tamped down a path in the deep snow with my snowshoes so I could drag the toboggan with a carboy on it, to collect the sap.

Now all is "running." It looks ridiculously simple. I'm just sitting back relaxing in the dusk before a crackling fire, staring into it with a relaxed, pleased expression as I slowly nurse a beer.

March 25
BIRCH SEEDS

The top of the snow now seems alive because there are tiny black springtails sprinkled like pepper all over it. They gather by the thousands in the footprints of deer, dogs, and people. And then there are the birch seeds. They are small grains with a little wing attached on each side, which helps them spread by the wind. The mechanical pounding of the rustling branches is disintegrating the drying cones, shedding the bracts, and releasing the seeds. I pick a dry cone from a birch tree, and it crumbles in my fingers, showering bracts and seeds upon the snow.

How many seeds are there in one cone? Gently I press another cone, releasing its minute cargo in little diffuse heaps over several square feet of snow. I count 350 seeds. The tree whose cone I picked has, on one branch, an average of about 115 cones, and there are 60 such branches on this medium-sized tree. In all, this one six-inch diameter gray birch still has about 2,415,000 seeds to shed, and it has already shed at least as many throughout the fall and winter until now, judging from the central cores or stems of the cones that still remained on the tree.

Gray birch twig, w. leaf buds, unopened new catkin, plus year-old "cones" — one unopened, one shedding seeds + bracts + 1 stem of shed cone

A production of nearly five million embryos for just one year, with possibly 20 years of lifetime production, gives this small tree a staggering output of 100 million embryos. All are created equal, or nearly so. But, on average, only one embryo of these 100 million will grow to become a mature, reproducing tree, assuming that the tree replaces itself.

There is a sharp line here between the potential and the actual. It's hard to have reverence for the life of one embryo when you consider its low relative worth in comparison to that of the adult tree, for which 99,999,999 embryos must die according to this species' design. On the other hand, it's all the easier to have reverence for *life*, when you see the complexity of the evolutionary design that produces this individual adult. Nature abhors the superfluous, yet is constrained to produce the seemingly extravagant.

March 30
COCOONS

The cocoon of a *Samia cecropia* moth that I found near the brook on a viburnum bush is a marvel of construction. With a sharp razor blade I force a cut through a papery layer of silk on the outside that was stronger than my briefcase (a Federal Express mail envelope), and this layer enclosed a secondary cocoon inside with the consistency and strength of hard leather. A layer of stiff silk threads separated the two. Even if a bird manages the unlikely task of tearing open the first layer of the cocoon, it must then face a great disappointment, for it now finds no juicy morsel but instead the hard ball of even tougher fiber. With sharp scissors, I am able to also cut open this secondary cocoon, finding inside a black and brittle, empty pupal skin. The moth had emerged from the cocoon the previous summer, and presumably it had done so much more easily than I had just gained entrance to it.

On the pupal skin that the moth left behind in the cocoon, I could still see the finely etched imprint of its abdominal segments, wings, eyes, antennae, and even its genitalia. It had been a male, judging both by the genitalia and by the broad, feathery form of the antennal casing. Squashed into the bottom, inside the cocoon and below the brittle pupal skin, was the rumpled cast-off caterpillar skin, the rudiment of the animal's *second* previous life. The caterpillar had probably fed for a month or more on the viburnum leaves before spinning its elaborate double cocoon. The caterpillar's skin had been soft, so when it was slipped off it had become compressed and shriveled into a ball at the bottom of the cocoon, like a wet shirt discarded into a corner. Only the hard thoracic leg casings and head capsule of the caterpillar still retained their form, like a pair of cast leggings and a helmet.

The fresh pupa, having cast off its pliable caterpillar skin, had hardened and become black-brown. It showed the future moth's features, however, long before there was any trace of moth inside. Not until the summer, after one full winter, had the moth finally formed. Then the brittle pupal case had cracked at the top, where the adult moth had emerged.

Before coming completely out of its hard case, the insect had defecated, leaving a fecal mass in the bottom. This mass, the meconium, consists primarily of solid white uric acid, the nitrogenous waste product of protein metabolism that has been produced throughout development. Unlike the urea that we excrete instead, uric acid can be excreted virtually without water; hence it is, in a sense, solid urine. Uric acid production had helped this animal to survive a year-long fast without one drink of water.

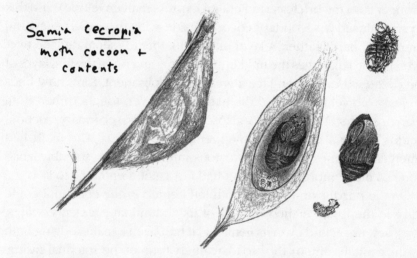

Samia cecropia moth cocoon + contents

The caterpillar must have pupated in late summer, two summers ago. Following the sub-zero temperatures of the winter before this one, the pupa's brain had measured the lengthening days and warmer temperatures of spring, and using this information as a signal, had initiated a precisely orchestrated cascade of various hormones that finally helped produce the moth inside the pupal case. Additional signals and the subsequent release of other hormones triggered and orchestrated the emergence behavior of the moth in early summer.

The pupal case had withstood the weather and possible attacks by chickadees and woodpeckers, and would appear to have been an impenetrable fortress. But at the end of the oblong cocoon where the pupa's head lay, the caterpillar spinning the cocoon had left an ingenious escape hatch to be used one (or, in some cases, several) years later when needed. There was also an escape hatch in the inner hard cocoon. Both escape hatches work like a sock with an elastic neck that can be opened by pushing out from the inside, but that closes shut if you attempt to push in from the outside.

The cocoon had survived two winters, as proven by the empty pupal skin of the moth, which was by now long dead. The adult does not survive the winter; it lives for only a few days after emerging in the summer. The moth's descendents were now already encased in other cocoons. Like many of us, the moth had by its constructions achieved an immortality of sorts. Immortality was nothing that *it* cared about. I wonder why we do.

I had picked the cecropia cocoon and brought it back to the cabin because it was a curiosity that I do not find often. Far more commonly I find the cocoons of the promethea moths, *Callosamia promethea*, in these winter woods. The handsome green caterpillars, with four bright red knobs on the front end, feed on ash in late summer; on these ash twigs you later also find the cocoons suspended, looking remarkably like just another stray, dead, curled-up leaf that has somehow hung on, like many withered leaves do all winter.

I returned to the young ash trees along the trail where I had seen a group of four promethea caterpillars last June (the issue of one female who happened to stop there to lay some eggs). There, I quickly found a cocoon because my search image is honed. As is typical of promethea cocoons, this one was rolled into one of the seven leaflets of a compound ash leaf. That leaflet and the central petiole of the leaf still remained on the twig, but you only saw the stubs of the other six leaflets where they had broken off in the fall.

The promethea cocoon has only a hint of the outer shell that is so prominent in the cecropia, but it is at least as tough, staying on the tree for several years. You'd expect that a cocoon rolled up into one little leaflet at the very end of a large compound leaf would inevitably fall onto the ground in autumn, when the leaflets, and finally the central petiole are shed. But no—the caterpillar has provided that this does not happen, and has arranged to stay above the ground through the winter. Looking closely along the old dried-up petiole you see a solid strap of silk, composed of perhaps hundreds of strands that the caterpillar must have laboriously laid down from its mouthparts (unlike a spider, which lays silk from spinnerets at the rear). The strap attaches to the cocoon at the bottom, and it wraps solidly round and round the branch at the top. The strongest winter gales cannot tear it off. Each time I see one of these cocoons hanging from a tree, all of these marvels flash through my mind. If they did not, my eyes would not see it at all.

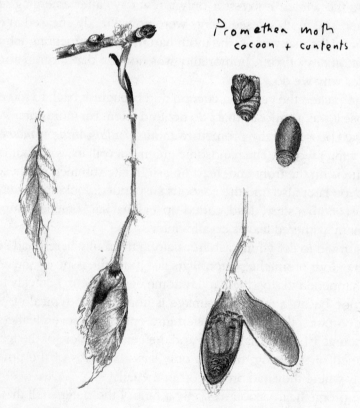

Promethea moth cocoon + contents

On this particular day I was in a cocoon-searching mood, and I found six of the promethea cocoons. In past years I had found many more. Were they becoming scarce? Coming home and cutting them open (because they seemed light and empty) I found a partial answer: four of them had been parasitized. One contained the larval skin of the caterpillar, but above it—where the moth pupa should have been—were 18 small, round, silken cocoons that had been spun by 18 small, parasitic wasp larvae. The cocoons were empty. Adult wasps must have escaped out the trapdoor at the top of the cocoon where the strap attaches.

The other three cocoons also had been parasitized by a wasp that must have laid her eggs into the caterpillar. But this invader was from a smaller species: she had used one pupa to feed 30 to 35 offspring rather than 18. Their many tiny cocoons now completely filled the moth cocoon.

Cecropia and promethea moths are members of the family *Saturniidae*, the giant silk moths. They are some of the largest and most spectacular moths that visit our porch lights in the summer. Two other species occur here quite regularly: the pale green luna moth and the polyphemus moth. Both have wondrous, large, luminescent green caterpillars that used to set my heart to pounding whenever I saw one as a boy. They still do.

The silken cocoons of these two other moths are tough roundish structures built inside leaves rolled together. In the fall, they are buried in the leaf litter. Unlike the cecropia and promethea moths, however, these two do *not* have built-in escape hatches for the emerging adults. Instead when they emerge, they secrete a saliva containing enzymes that digest the tough silk protein of the cocoon, and so it is that the moths make their own escape hatch. I do not know why the first two species attempt to stay off the ground, while these two don't, or why there are such totally different mechanisms for escaping the cocoon, all in the same family of moths. As in all living things, each creature does it its own way, and very small, perhaps almost random selective pressures may start a cascade in one direction that, once started, continues on its own momentum to ever greater differentiation and perfection. Not a shred of intelligence is needed to make it, nor to make it work. It is strange to me that life itself doesn't strike the average person as all that impressive, but for some reason "intelligent" life does. When you consider life as a whole, intelligence is a mere bristle on the hog.

I kept my eye out for silkmoth cocoons—in all I found 33. Three were of the cecropia moth and 30 of the promethea moth. Not one held a current live moth pupa, although 15 contained pupal skins showing that moths had successfully emerged at one time. Many of the pupal cases were, however, already several years old. For the 18 that had succumbed, I determined various causes of death. Two each contained one single large, round, pill-shaped ichneumon wasp pupa. Eight each had been parasitized by a wasp making 20 to 30 offspring in the cocoon. Four had been parasitized by fly larvae that pupated within the caterpillar, which then died and never molted to the pupa. One caterpillar was mushy, presumably the victim of a virus or bacterial infection. One was overgrown with fungus. Finally, two very thin cocoons had been torn open and the contents removed. I suspect a bird had found and eaten the caterpillars before they were finished spinning their protective shell. By contrast, all of the parasites gained access to the larvae *before* they spun cocoons.

I tried to see the "faces" of the, to me, unknown parasitic wasps locked into the cocoons, but I was unaware of the special requirements for keeping them alive. The two large pill-shaped pupae died, and only two half-centimeter-long female cryptinae ichneumon wasps with even longer antennae emerged on August 15 of the next year. They were beautiful creatures with red legs, black head and thorax, and black white-ringed antennae. Their abdomen was red at the front with black further back, but the end of it was tipped in white and terminated in a two-millimeter-long ovipositor.

Despite the "perfection" of the moths, at least six different other organisms have their own perfections with which to surmount those of their prey. Should the caterpillars ever threaten to become numerous, any and all of their natural enemies will assuredly multiply and catch up quickly to make use of the new resources. In the long run, therefore, like all other creatures, the beautiful silk moths are necessarily rare at times, though nevertheless just as solid and necessary a part of the fabric of the forest.

During the next fall and winter I searched again for the saturniid cocoons. I did not find a single one. I'm confident that it will be some years before I do.

March 31

More flies. In the last three days I've sucked up 1,600; 900; and 400, respectively. Maybe I'm making headway, although it seemed last fall that I'd gotten them all, too. These few are the remaining overwintering guests who are just waking up. The rest will be rushing out of hiding in the cracks and crannies soon, and the cabin will then be clear of them until fall.

Reawakening life is all around. Each insect overwinters in its own specific life-cycle stage, and eggs, larvae, and pupae all are becoming activated. The first birds are singing: I heard a mourning dove and a purple finch near the cabin this morning. The barred owl, a very close cousin to the famous (or infamous) spotted owl, now calls every night.

Dry tan beech leaves that have survived all the winter storms now rattle softly on a low branch. An equally joyous new sound is the rushing and roaring of the brook. Where there was white silence two days ago, there is now a raging yellow-green torrent flowing over the remaining ice. The three warm days that awoke the flies in the house have also considerably lowered the snow level. Now it freezes at night into a crust that you can walk on in the morning. And, sure enough, the snow's surface is no longer immaculately white, but is studded with seeds and debris blown from the trees.

Each seed contains a tree embryo, the *information* to make a tree, which can become expressed if it becomes implanted in the earthen womb by an umbilical cord called a root. A tree's life is an extraordinary achievement against incredible odds, from an individualistic perspective. From the perspective of nature, on the other hand, there is assurance that each tree will produce, on the average, just one other reproducing tree. How different from our current society's perspective, where every life has a "right" and where, if any individual does not have ideal conditions in which to grow, we look to find who is at fault. We have ideas about how life is "supposed" to be. But to a tree, and to most other organisms, life itself is the very ideal of the "luck of the draw." Luck overwhelmingly accounts for all success. Individual differences matter—but most are born equal.

The world we inhabit is built on chaos rather than on a predetermined order. And that is precisely what I find to be uplifting, and food for joyful optimism. Tocqueville said, "Chance does nothing that has not been prepared beforehand." Perhaps I had prepared for being here, by dreaming it as a boy.

I turned away from the brook and felt strangely restless. I had not been prepared to confront the ghosts of the past—ghosts that had long ago tormented me when I ran away from school to live in the woods.

Without much thought I drove to the diner, to read the paper in the company of others, as well as to eat bacon and eggs and drink coffee. I needed to get away from my own cooking.

April 2
DREAMS

I had thought spring was here because the snow had been melting all last week. The first patches of bare ground showed around some of the larger tree trunks, and the brook ran amber over the ice. But yesterday we got fooled—a blizzard raged all day long, and all night long the wind howled.

In the morning yesterday, just as the storm started, I got the urge to see three of the local raven nests. At all three the female was on the nest, and the male perched nearby in the trees. Maybe I was fooled about spring, but they weren't.

I spent the day doing a watercolor of twigs of bog shrubs and trees with flower buds. Maybe the Cro-Magnon cave people were dreaming of the animals they hoped to see when they drew them on the cave

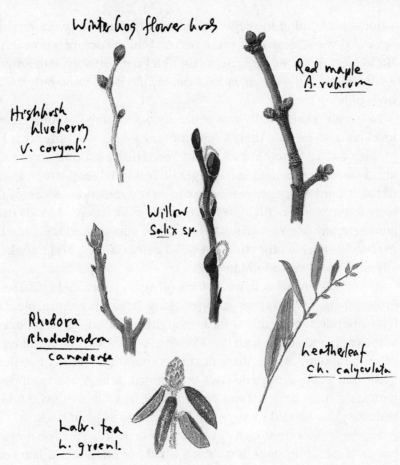

Winter bog flower buds

Red maple
A. rubrum

Highbush
blueberry
V. corymb.

Willow
Salix sp.

Rhodora
Rhododendron
canadense

Leatherleaf
Ch. calyculata

Labr. tea
L. greenl.

walls. I'm dreaming of spring flowers, and I'm drawing the buds in the cold, dark confines of my own cave, as if drawing them will make the flowers appear.

Even when awake in the daytime, I have a recurrent memory of a dream I had a few days ago. I do not remember all the details. But I vividly remember the owl. It was a great horned owl. It was looking at me and I marveled at how very large it was. I knew it was a female, because in owls they are larger than males. Her mate was nearby, but mostly hidden in the trees. Also, her color was unusual—slightly grayer than Bubo, the tame owl that I once had roaming in these woods, yet coming to me when called. I had always wondered what sex Bubo was. Comparing these two owls in my dream, I now thought Bubo must have been a male, as though I had just solved the mystery. As the edges of the dream faded from my memory, I saw only the large owl studying me.

Last night I dreamed I was aimlessly wandering, looking for a dry level place to sleep. I do not remember where I was, only what I was looking for, as if nothing else mattered. Finally I found level ground and a place that seemed like it might be comfortable. But then I saw, in terror, a herd of elephants galloping in my direction. Some of them were angry. Some of them were wounded, with tattered bloody trunks. Strangely, my fear dissipated even as they surrounded me. One lifted me onto its back with its trunk. It was a gentle animal, and it spoke like a human in a deep soothing voice.

After I was on its back, we all took off again on a run. I could see the ground far below me as we galloped along. It felt smooth and pleasant. I felt suspended above the earth, as though on a cloud, yet still connected to the earth and at one with these beasts. As if I wasn't being taken care of well enough, *my* elephant next lifted me onto another even larger beast that came alongside. But this animal, which at first appeared menacing, turned out to be as benign as the first. On we flew. And I was in ecstasy, for I could survey even farther from its great back.

Suddenly, incongruously, I was on the ground again and there was not an elephant in sight. It was as if I had forgotten them. They could have been extinct. The ground now was paved with meat, mostly the linings of stomachs, which are white. There was no *blood*, as such. There were faceless people around who were doing something half-heartedly with this meat. The only displeasure I felt was in not seeing a dry level place to lie down and make myself comfortable, given all the meat lying around. Not seeing a suitable bare spot close by, I started running to try to find one. I ran on and on. But all was the same, wherever I went.

Eventually I met three dogs. They were big and very gentle. I was again aware of faceless people with these dogs, but the humans were hazy—I hardly saw them. The people treated these dogs kindly, fed them all the meat they wanted. The dogs produced litter after litter of puppies, and in time there was a snarling mass of vicious dogs. Suddenly alarmed, the people who had dearly loved the dogs were forced to hold their puppies under water till they drowned.

Then I woke up. I knew I was awake because I struck a match, lit the oil lamp, and wrote down what I had just seen. It was 3:34 AM. The wind howled, and my feet were cold. Indian warriors used to go out alone in the wilderness to have visions. Was I beginning to have mine?

I got up, rekindled the fire, and then went back to bed and slept until seven. I awoke to rain and sleet pounding on the roof, and to the sound of rushing wind. My feet had warmed up, so I had slept more soundly, without dreaming.

As I opened the cabin door to dip my kettle into the snow to get water for my morning coffee, I realized how much warmer the air had become. In the drizzle and the fog over the forest, I saw two ravens gliding along. They flew side by side, making large, irregular circles. They were playing with the breeze, occasionally letting themselves be lifted up, then folding their wings and diving down side by side, like black bolts out of the sky.

April 3
ICE FISHING

The Ridge Road runs south from the one-store village of Chesterville, on top of a glacial esker. Horseshoe Pond, Round Pond, and Norcross Pond lie to the west of it; Fellows Pond lies to the east. All are embedded in a continuous stretch of spruce bog and a low woodland of pine, birch, and maple, with red oaks higher up on the esker itself. As you continue south you pass a few summer camps along the ponds and then, as you leave the esker, the road is unpaved. In early April it becomes a quagmire where people challenge their four-wheel-drives in the mud. Bill Adams and I went ice fishing at Mosher Pond, two or three miles below the esker. It's just before you get to Twelve Corners, where the map prominently shows roads from six directions converging to one crossroad. In reality, the roads look considerably less prominent on the ground.

We stopped under a snowbank along the road by the pond, and pitched our gear out onto it, climbed up, and then on snowshoes clambered down the steep slope to near the outlet, where Bill felt the fishing would be the best.

The pond was covered with an old snow crust, almost but not quite firm enough to walk on with snowshoes. Underneath the crust and on top of the three-foot-thick solid pond ice was a foot and a half of ice-water snow slurry. It made for hazardous walking, unless you didn't mind your boots full of ice water. To chop a hole through the solid ice underneath was also a challenge, but Bill had brought along a Swedish ice auger.

After half an hour or so of vigorous boring with the hand drill, we both worked up a sweat. We had even penetrated the ice. Unfortunately we'd struck a shallow part of the pond—the bottom of the ice layer was less than six inches above the mud. Not a good place to catch big fish. If we were each going to set our five traps, we'd have a full afternoon, and by now it was already noon.

Another ice fisherman with a small boy came buzzing over on a snowmobile from the other side of the pond. "Get any?" we asked him.

"A few," he replied, and stood by his snowmobile watching us try to drill a hole. "I've got a power auger—want me to drill some holes for you?" he offered.

Bill, who was sweating and straining, acted hesitant. After all, he didn't want the peace and tranquility disturbed by the noise of a gasoline motor. But facts are stubborn.

"Ah . . . well, sure, we'll take a couple."

Zip. Zip. Two holes in the ice. That's more like it.

"Want some more?"

Before Bill had a chance to refuse, I said, "Yes—about *ten* of 'em." And I marked out the spots.

In five minutes it was done. The fisherman left and the pond was again quiet as a church on Tuesday afternoon.

We set our ten traps, each rigged so that a spring releases a red flag that pops up when a fish grabs our minnow on the hook. Then we brought over our aluminum lawn chairs and a couple of six-packs of beer, to soak up the warm spring sunshine.

Heat waves were shimmering over the shadows of the pines on the ice covering the east shore. We chewed salted sunflower seeds, and slaked our thirst. We sat and relaxed for an hour, or two. No red flag went up.

"They'll start to bite in an hour or two. It's too early yet," Bill said. An hour or two later. Still no flag up.

All entertainment is slow in Maine. If you have to pull up little pickerel for entertainment, rather than large salmon, then pulling up pickerel is your ecstasy. There were no salmon in this pond. We were expecting little pickerel. But where were they?

Once in a while someone drove by down the dirt road along the shore. They'd blow their horn, just so we could see them wave hi to us.

Another hour went by. Still no flag had gone up to signal that a pickerel had grabbed our bait.

"This is *fishing*," Bill said after a while. "It's just like going hunting— you never get anything. If you always *got* something, you'd say you were 'going killing.' "

After another can or two, which Bill kept cool by sticking in the ice slush at his feet, we started reminiscing.

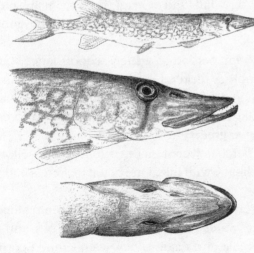

14" pickerel ♀ - tadpole in stomach much yellow roe

- stripe under eye to reduce glare?

top is olive green (sides yellowish + belly white)

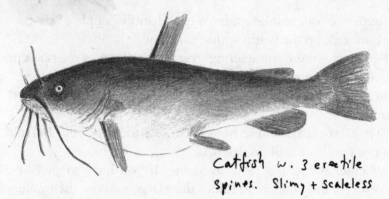

Catfish w. 3 erectile spines. Slimy + scaleless

The sun was moving across the sky and we had almost forgotten to check our traps. But Bill looked up and yelled, "Tuna!" then pulled up a line to land a ten-inch chain pickerel. Close enough. You could see why they are called "chain" pickerel, seeing the lattice of chainlike gray markings along its dark green top and yellowish sides and belly.

Eventually we even caught a hornpout, known also as a catfish because of the long whiskery tentacles on its mouth. This soft-bellied fish is not as defenseless as it seems. Pulled out of water, it immediately locked two needle-sharp spines into place that must discourage the hungriest heron from trying to swallow it. The two spines are the first rays of each pectoral fin. To move the fin the spines have to be moveable; these assuredly were not. What kind of magic does the fish have, that it could have a *moveable* fin that is a stiff pincushion at the same time?

Bill had to leave early to chaperone at a high school dance, but before we left we caught another pickerel, a silvery pond shiner, and two yellow perch with bright red on the fin rays and on the second pair of ventral and anal fins. Like the pickerel, the perch were colored beautifully in yellow and olive green.

Both pickerels and both perches were females swollen with yellow roe. The shiner was a male, heavy with milt. One of the pickerels had a tadpole in its stomach.

I sketched the fish to preserve their images in my mind, then scalded and gutted them in boiling water. The skin easily came off, and I removed the tender white meat and made a stew with canned corn, broccoli, and milk. With the catfish, I even enjoyed the bones. I boiled the meat off the pelvic girdle and studied the mechanism whereby the fish could lock its spines in place and then again unlock them so it could swim. I wondered if, like the inventor of Velcro who must have gotten the idea from a seed of burdock, I could find an application for what the fish had invented. Perhaps unlike Thoreau I *am* interested in "the guts of a hawk"—maybe even those of a fish. There is beauty not only in that things work, but *how* they work.

April 4
RETURNS

Springtime *has* come to Maine. It's peak sugaring time. I boiled sap again last night under the full moon. Some think the full moon makes the sap run, and run it did all day. I hauled in four full five-gallon carboys from my 20 taps.

The night air was crystal clear and chilly. The moon's neon glow illuminated the snow in milky white. I stared at the moon, seeing its clearly outlined gray lunar seas. The hills and woods were black in the distance all around, even as the snow underneath the young maple trees near me shone crisscrossed by the moon shadows of the trees. The black silhouettes of the maples showed against the sky near the Big Dipper, almost overhead. Orion was far to the west, whereas all through the winter I'd been seeing him more toward the east.

Smelling the sweet aroma of the maple sap and squinting as I sat next to the fire under the evaporator, I was reading Alan Lightman's *Ancient Light*, about the creation of the universe. They've hypothesized what happened between when the universe was "much smaller than the nucleus of an atom" until an infinitesimal fraction of a second later, when it already measured 10^{400} light-years across (that's a 10 with 400 zeros behind it). Sure.

By the time I'd had my second beer, I decided that what the physicists have done, really, is to destroy the validity of common sense—of believing in the rock we stand on. As for me, I still believe in trees.

Meanwhile, I'm warmed next to the fire in the rusted barrel stove. Orange light flickers through its holes onto the woodpile behind me, and billowing clouds of steam rise off the syrup into the clear sky. I'm mesmerized by the tranquility. Tomorrow I'll bottle the syrup.

Eventually I got two gallons. So I really *did* get something accomplished by sitting in the moonlight. It feels a little bit like having a hive of bees working to make honey for you. There is a lot of effort involved, but you know you couldn't do it yourself, gathering the honey from all those flowers. Here I'm letting the sun, the trees, and the fire work for me.

This morning is another glorious one. I sit outside to watch the sun come up by Kinney's Head. As usual, I stare at the hills, as I've done for perhaps hundreds of hours by now. Just before the sun comes up, the barred owl hoots in the woods behind me, and at nearly the same time, the ravens wake up and call. The sun creeps up the mountain and warms the cabin's gray logs. Flies buzz. Then—a lone red-winged blackbird flies over! And then a tree swallow! Miracles—the first migrants are back. Both call continuously and excitedly. Then I hear a robin scold from the forest.

It will be weeks before the groundhog comes up. Woodchucks feed on greens, and the snow has to melt before they grow. In Maine, people gauge whether spring has arrived not by groundhogs, but by skunks. Skunks roll out of their dens in February or March and embark on a long prowl. Six to eight young are born two months later.

In the *Waterville Sentinel,* Joe Rankin writes that the skunks that had spent the winter beneath a portable classroom at Sidney's Bean School have been making life difficult for students and teachers alike in the last few weeks. However, the faint periodic odors weren't nearly so objectionable until one of the skunks contested a custodian's attempt to relocate it.

After a leisurely breakfast, I feel fortified again with plans, with ideas, and with purpose. I drive out to find dead calves. I open the windows of my truck wide, to smell the fresh clean air. All along the route I pass old farmsteads surrounded by ancient sugar maple trees that dangle plastic milk containers. Everyone is cashing in on the chance to make a little syrup.

A farmer just had a calf die, so I'm in luck. I tell him that I've seen fewer ravens now in the last several weeks, and ask whether that is his impression, too.

"I see them all summer. But in the last week I've seen vultures for the first time. They're never here in the winter. Also," he tells me, "I saw a great big black bird with a pure white tail." I guess he didn't notice the head.

He wants to know if I'm working "for" someone, or just doing this on my own. It's a tough call, even though I've been asked this seemingly most absorbing question many times before.

On my way home I saw three ravens in a backyard at a house along the roadside. There was a doghouse near where they stood on the snow, and I wondered if they were trying to steal the dog's food. The ravens continued to loiter, then flew into the woods. I drove on slowly, but a little further ahead I saw them come out of the woods and land along the side of the road. They sauntered along as if on parade, occasionally picking at the gravel. They reminded me of three little schoolboys I had just seen earlier this morning, waiting for the school bus.

There was no traffic and I stopped the truck, watching the birds for about five minutes. Eventually a car came from the other direction, and the birds again flew into the nearby woods. After a minute or two, a pair of them flew up above the trees and started to circle. The third one joined the other two, and the three then circled in a tight cluster for well over ten minutes. Higher and higher they went, alternately flapping and soaring. All the while they kept up an almost constant chatter of low grunts, quorks, and knocking sounds. When they were so high that they were only black specks to my eyes, they suddenly departed in a straight line, flying due north, where they disappeared out of my vision over the horizon.

April 8

Paul Melcher is a burly man, 78 years of age with a gray crew cut, who operates a nearby small dairy farm with his wife, Laura. Little has changed for the past 45 years or so since they first settled there. The farmstead is at the end of a dirt road, and by looking at the rusted and decaying automobiles and farm machinery strewn around the fields, you'd think all the traffic on that road has been one-way for some time. The fields covering the gentle hill extend down to a brook that borders a steeply wooded slope with hemlocks and pines, where I hunted deer as a kid.

I'm here today to check for a possible raven nest. Paul ambles out of the shed, .22 rifle in hand. He has a red squirrel problem.

"Get any?" I ask.

"I got one. But I think four came to the funeral. These damn squirrels are worse than rats once they get in your house. It's either me or them. And I'm afraid it's not going to be me." (Red squirrels are notorious for chewing old wiring and generating sparks on dry tinder.)

We chat on about ravens and before I go across the field he says, "Wait a minute." He goes in the house and brings out two big yellow apples and two molasses cookies. "Here. And when you come back, tell me what you found. I'll be inside the barn, milking."

Walking down the hill, I come first to the big flat field bordered by the brook. There already are several bare spots where the sun has burned through the snow to the ground. Two killdeer have just returned and found this ground, and are calling loudly and exuberantly from it. Out of the speckled-alder thicket at the edge of the field, I also flush the season's first woodcock. The male alder catkins have transformed overnight from tight little purplish brown sausages to long, loosely dangling ornaments that when touched shed puffs of bright yellow pollen. The female hazelnut flowers have extended their tiny purple tongues, but the male flowers resemble those of the alders.

I walk up the hill, following Paul's tractor tracks. He has just dragged out the pine logs he had cut in the winter, and I find the raven nest in the same tree it had occupied four years ago, close to where Paul has cut the logs. A goshawk nest is within 15 paces of the raven nest. As I get close the raven flies away, but the goshawk warns me with its loud, rattling "kek-kek-kek" calls. Then it launches itself off a tall pine, aiming directly for my head and coming within less than a foot before veering up sharply.

When I get back, Paul is in the barn, milking his Guernseys. A red squirrel is chattering in the shed. The inside of his barn seems comfortable and friendly, like the old Adams farm. It has windows. The cows have plenty of bedding. Most of the Guernseys are hitched to their stalls, but one is in a special stall. She has two young calves that look remarkably like fawns without spots. English sparrows are chirping on the rafters and underfoot. Paul doesn't know what kind of birds they are, but says they've spent all winter there in the barn.

"I like them!" I confess.

"The milk inspector doesn't!" he says. "We've been trying to get rid of them, but haven't succeeded."

I tell him that I've found the raven nest and that it has five speckled blue-green eggs.

"Was it near where I was cutting?" I tell him it is. "Well, there are plenty more trees. Why don't you mark the tree so's I don't cut it."

Each of the 25 pine logs Paul has cut is worth at least $75. And I think about how Paul does "forestry." He has no degree in it. He's a farmer. But his woods look as beautiful as any tended by a forester with a postgraduate degree from Yale. They are home not only to a pair of nesting ravens, but a goshawk as well. There are grouse, deer, fox, coyote, and porcupine. His woods seem full of large and small trees, of many species. Whenever he wants cash, he drives into his woods with his tractor, and brings out 10 or 20 big trees. And he has been doing this at least since I was a kid on the neighboring farm.

The Boise Cascade forester came at 7 AM, and we walked my woods to check the cut. Later in the day he would go to the mill to work on the computer, then on to Augusta to testify in some tax hearings, to seek tax relief to help cover forest management costs. Spraying ain't cheap.

The crust was hard and we walked freely. That was a treat. Everywhere there were deer tracks made while the snow was still soft. We judged that the loggers had done a good job. I've got more money in my pocket and the loggers have, too. And I feel good about my forest. It is being restored to a variable-age stand from a thicket that grew up in fields. The big trees left will grow even bigger and faster now, and the light that gets down will now regenerate young trees.

When I got back to the cabin the sky had started to cloud over. It might rain soon. The lone tree swallow, perhaps the same one I had seen for the last two days, was swirling around in the clearing. A flock of 15 robins came by briefly, then moved on. The first phoebes and song sparrows returned today.

Thousands of little flies from a species that never comes indoors were buzzing all over outside of the cabin. They sounded like bees.

I saw two black arctiid caterpillars wandering far out on the ice in the center of Hills Pond, and a very tattered Compton's tortoiseshell butterfly settled on a snowbank along the road.

With the snowpack quickly receding, I now saw that all of my planted apple trees, as well as many small maple trees, had had all of their bark removed near the ground where they had been under snow. The mice had been active in the clearing under the snow cover, accomplishing what I tried to do all summer long with bush-cutters, and much more. In the forest, however, *none* of the small trees had been girdled by rodents, because there was none of the matted grass that the meadow voles need. Until I noticed this phenomenon, had I designed a computer program to manage my meadow, I would never have noticed the mice, nor even one of the many thousands of wasp species that parasitize caterpillars. But both of these, and a thousand more variables, make all the difference in the forest that runs itself quite beautifully.

Tattered Compton's tortoise shell –
just out of hibern. On snowbk.
Half of rt. lower wing missing.

Curled up (after touching)
woolly bear caterpillar –
2 found wandering on
ice ~ 200 yds fm shore
on lake

Winterberry – on southf. slope.
Red berries from last year!
Taste like spearmint – good!
Thick leathery green leaves.

April 11

It rained all day yesterday, all night, and it was still raining this morning when nine robins came by. One sang. They ran over the matted grass and pecked at the rosehips that remain on my wild rose bushes from last summer. One gave a loud call and then they all flew off as a group. When the downpour let up I heard crows cawing, and a purple finch and a junco.

The ruffed grouse, as I saw on a roadkill, are discarding their "snowshoes."

April 20

A winter wren in dense brush made the rapid churring "trrrrr . . ."
Another one sang its vibrant melody in the distance. While in the
woods, I also heard several brown creepers singing, and the first sad,
melodic refrain of a hermit thrush again burst forth in the evening. The
sapsucker too is back, calling and drumming.

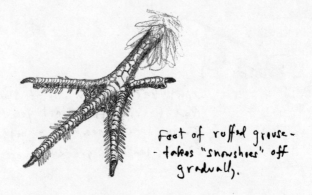

Foot of ruffed grouse -
- takes "snowshoes" off
gradually.

The song of the brown creeper is short, but it is one of the most
impressive songs around, with an incredible sequence of notes
crammed into not more than a second. To appreciate all of this beauty
relies upon your noticing its subtleties. To hear them, you have to slow
down—way down.

I may not have really noticed it, but I think that since "growing up"
I'd been constantly taking on more and more responsibilities and
projects. To have any chance of getting them done, I'd gradually been
speeding up until finally I arrived in the fast lane. The landscape had
become a blur; after a while, I did not really see it any more. I may
still have known it was there, but only because I recognized the cues
from previous experience. Perhaps life was flashing by like a tape
player speeded up. I recognized the sound, but I did not hear the
music.

I've spent every single day this year looking at the hills. I've spent hours doing "nothing." Days. Maybe weeks. I do not think it has been wasted.

April 21

Stuart and I went for a long walk in the woods. I let him lead, to encourage his initiative and sense of discovery. But I did point out to him interesting things that we came upon.

We were walking in the maple grove next to a fir thicket, when out of the firs came the low "thump . . . thump . . . thump . . . thump . . . thump . . . ," ever faster, until it ended in a flutter. "Grouse drumming!"

We also heard the twitter of purple finches singing in the tops of the fir trees, and the songs of both golden-crowned and ruby-crowned kinglets. We rested in some old-growth conifers where these birds sang, and Stuart got excited about the different kinds of moss after I showed him a couple of them.

In the evening he watched the elaborate display of the woodcock.

"Let's see who is the first to see it fly up, and who can follow it the longest." It was dusk then.

He saw it take off first. "See it there go higher and higher?" The bird rose whistling like a giant hummingbird, circling in the sky above the cabin and the clearing until it was a speck in the darkening sky.

"That sound is made by its wings," I told him.

The bird was coming down. "Now it's falling like a leaf," he said. "Is that so it won't be recognized by predators?"

"No, I think it just *dives* down, and it just *looks* like a leaf falling because it's coming down. But it's colored like leaves, so you don't see it when it hides on the *ground*, especially when sitting on the eggs."

The bird was chittering, reaching the climax of its display, then fluttering back down onto a patch of matted grass amongst the spirea bushes. Stuart ran to it, flushing it, and the bird rose again to repeat the display.

*   *   *

The hillsides are still gray from the bare twigs. But every day we see more pastel patches of red, purple, yellow, and pale green of swelling buds. The sight is not spectacular. Nevertheless, its subtle beauty is assured by the long months of winter when the snow swirled through the trees and only the hardiest animals stayed in a world of white birches, gray maples, and black-green spruces. Now millions more birds are on their way back again. They'll be here in days to begin their jubilation. Right now, spring is my favorite season. In a month the indigo bunting will sing and build its nest in the brambles. Two months later, the maples will turn red and yellow. That will again be my favorite season. Here, my favorite season is always the one I'm in.

April 28

There are still nightly frosts, but in the mornings the birds sing. Right now outside the cabin at 8 AM a flicker cackles, and at least three pairs of tree swallows gurgle and inspect the bird boxes I put up. The robin sings its glorious song. Throughout the day the grouse drums in the woods, and the woodcock performs its exuberant ritual at dawn and dusk. A purple finch sings ebulliently for hours. What a surprise—it is *not* colored purple like a male, but brown like a female.

While in Vermont to take Stuart back to his mother, I saw the first delicate spring flowers that had popped like magic from the forest's damp leaf mold. The blue and white hepaticas were in full bloom, shining like patches of jewels, as were the pink spring beauties. Bloodroot, Dutchman's breeches, trilliums, and trout lilies were unfurling their distinctive showy leaves, but their flowers had not yet opened. A sedge was ablaze with yellow spikes. Here on the hill it is still too early for the flower show on the forest floor, though the woods already have a slight red tinge from the blossoming of the red maples.

April 28
FREE SEARCH

The ice left Hills Pond two or three days ago, and fishermen now stop there daily to cast lines for trout. The loons are not yet back, but they should be here in a couple of days.

The woods are in color. Most of the wind-pollinated trees—quaking aspen, balsam poplar, bigtooth aspen, speckled alder, willow, hazelnut, elm, boxelder, and red maple—are in full bloom. As in the fall, the most striking display is again provided by the red maples. Red maple twigs are reddish purple at the tips and gray further down, and whole hillsides that seemed dull just a few days ago are now awash with bright red, yellow, orange, and deep crimson.

I'm surprised at the range of color in the flowers of one species. The red maples probably are pollinated by both the wind and insects. The wind doesn't care what color the flowers are, but individual insects do. Insects become conditioned to a particular color, and after having received a nectar reward at yellow, for example, they'll search for other yellow flowers. So it pays for all the flowers of any single species to have the same color in order to promote cross-pollination. So why do different red maple trees have differently colored flowers? Might a *tree* in a densely growing stand have so many flowers as to keep a bee from visiting distant, unrelated, trees for cross-pollination? If so, then the greater the color variety among individual trees, then the further the bee could be induced to fly if it always visited similarly colored flowers. However, the reality turns out to be a bit more complex: I threw up sticks and knocked down a few flowers, discovering that all the trees with deep purple flowers had only female flowers. The male trees had reddish unopened flowers that reveal yellow as they open.

I went into the woods without looking for anything in particular, and without any predetermined goal. At first my legs took me down the old logging trail. Examining a freshly cut, solid red maple stump of over a foot in diameter, I saw scattered through it eleven round quarter-inch holes. They were drilled straight down. I poked each of them with a twig, and they all went at least a foot down toward the roots. There was no way of knowing how far *up* they had gone, because the log was gone. What grub had chewed galleries in solid maple wood? And why were there so many in one tree? What factors had prevented the insect from multiplying and taking *all* the red maples?

For a while I tried to follow a kinglet, hoping it might lead me to its nest. But then I realized it could enter its nest a dozen times in the thick tangles of a spruce and I still wouldn't see the nest. So I went on, down to the brook. On the opposite shore I saw two large gray black shapes: moose! Their hair was in raggedy patches, for they were molting their thick winter coats. As I drew close they both bolted, crashing loudly through the alder thicket.

I was pleased to find some recently debarked twigs in the brook. These meant that a beaver had been here recently. Someone trapped them out three years ago at my swimming hole, and I'd been hoping the beavers would reestablish themselves.

A mourning cloak butterfly flew up from a tree trunk in the sunshine where it was basking. I also saw two comma butterflies. They fluttered through the thick shaded woods, occasionally stopping to bask.

The early warblers are back. I saw and heard several myrtle warblers, along with the first pine, black-and-white, and Nashville warblers. All of these were rare pioneers this year, and all sang only sporadically, except for the pine warbler. It stayed high up in the tallest pines, singing first on one tree, then quickly flying a hundred yards away to sing on a distant one, to then move on again. It seemed unsettled, as if trying to cover a lot of territory, possibly looking for a mate. A myrtle warbler in a flowering maple tree repeatedly probed into the massed flowers. Was it feeding on nectar? Its sibilant, almost whispering song lent a delicacy to the woods that contrasted with the exuberant refrain of a winter wren, close to the ground in some dense pine slash from the winter's logging operation. Ruby-crowned kinglets called in a peripatetic twitter somewhere in the distance, punctuated by the quick, sweet refrain of a brown creeper.

A red-breasted nuthatch crouched and fluttered on a branch. Then I saw her mate. I knew then from their behavior that their nest was near, and that it would now or soon have fresh eggs. So I stayed and watched to find the nest. The male sat on a branch making long-drawn, nasal calls, and I noticed that unlike most other birds, while singing he kept his bill closed and his throat puffed out like an owl's. Maybe that is why the call has that nasal quality. After several minutes the bird flew to a decaying alder stump, and perched upside down over a freshly-excavated hole that had some pitch plastered around the entrance. It peeked in, but then flew away. I had found the nest.

I now was drawn on to the distance by a chorus of wood frogs. When I came within 30 feet of their pool, its glassy surface was cast into a roil as dozens of places erupted in ripples. At the same instant, the ducklike chorus of quacking-croaking calls stopped totally. Slowly I came closer, walking quietly over the damp leaves. I peered onto the bottom of the shallow pool, but I saw no frogs. The bottom was lined with a layer of brown maple leaves from last fall. Somewhere underneath them, the tan and brown frogs with their black eye-stripe would be hiding. But dozens of round, globular masses of jelly full of black spots, the eggs, already were anchored on a birch branch on one side of the pool. I sat near the pool for ten minutes, but nothing stirred. In two months, thousands of little froglets would leave this drying pool and disperse into the woods.

My next stop, at the gravel pit, was on the way to another promising stand of white pines where ravens might eventually nest (there had never been a nest there before.) Here the sand bank erodes a foot or more every winter, revealing the feather-lined grass and pine-needle nests of the bank swallows. The swallows dig their burrows here in the sand every year. They had not come back yet. But a kingfisher probably had been back for some time, since it had already dug its nest burrow near the bank-swallow colony. I saw two fresh grooves side by side leading into the hole, made by their feet. Unlike birds that run on the ground, kingfishers have very short legs and they step with one foot beside the other rather than putting one foot in front of the other.

On the top of the bank there was a scattering of hairycap moss, lichens, stunted birch seedlings, and shriveled rosettes of last year's paintbrush flowers. Andrenid bees—almost all males—were zigzagging an inch or two above the ground. There were no flowers here, so they could only have been searching for females. Looking up close, I saw entrance burrows here and there that these solitary bees had dug, and in some of them a bee was perched inside near the top at the entrance.

There are no more flies in my house because those that made it through the winter left by the cracks they had entered in the fall. I am glad to be rid of them, but the thousands of species outdoors delight me. The first ones could now even be heard. In the cool, coniferous woods in places where the sun reaches down, I could make out a high-toned buzz like a miniature power saw. It was the sound of syrphid flies. This species has handsomely black with yellow stripes. They darted with lightning speed after one another. Then they would re-alight onto a spruce branch to renew their whine, the auditory record of their shivering to keep warm.

In my walk through the deciduous woods I kicked open a moss-covered log revealing galleries of big carpenter ants in the still solid heartwood. Unlike those I had found in the winter, these ants crawled about weakly. In the winter, it had taken several days at room temperature before they showed any signs of life. Their antifreeze was already gone, and they now had a pleasant nutty flavor that I thought I could get to like.

Many bumblebees flew a zigzag course close to the ground, looking for mouse holes. Sometimes they also flew high, fast, and straight as if they had appointments at distant destinations.

At 6:30 PM when I returned to my clearing, I started to follow a queen bumblebee. I followed her for 25 minutes. During all of this time she flew almost continuously, but I traveled only 20 paces. She hovered back and forth just barely above the ground, often returning to the same area again and again. Every couple of minutes she landed on the ground in the matted dead grass searching for mouse holes, then again resumed her flight. The only "real" hole in the ground I saw her enter was a posthole from one of the sticks I had used for last summer's beans. She came out in a hurry; then I lost her.

I've seen my first "blue"—a pale, pure blue butterfly that flutters through the forest on sunny days. Big black mosquitos are now also already active. These can be quite annoying at night, but their hum is worse than their bite. As you try to fend them off in the dark they readily retreat, then patiently wait a few minutes before advancing once again. As with many other minor annoyances (but unlike with potential problems), the best way to deal with them is to ignore them until they bite.

May 2

Bog Stream in Chesterville has always been a magical place, ever since Phil took me there as a small boy to fish, to teach me how to handle a canoe, and to see wildlife. As then, my canoe ascends noiselessly over the black, glassy surface. The dark, vegetation-stained water runs swiftly but smoothly, making tiny standing ripples that wind slowly like random lines over its surface. This early in the year, the water still washes over the shallow banks of the winding stream, flooding the marsh. In the matted, dry yellow grass, short new green spikes poke through, standing straight, erect, and sharp. Except for the red osier dogwood growing along the bank and reflected onto the water near the shore, the predominant colors are muted pastels. Mallards whistle overhead, and I envision the brilliant metallic greens of their heads and the blue specula of their wings, which look black at a distance. Around a bend, a pair of wood ducks makes squeaking calls and rises with splashing and then whistling wings. You do not see the male's brilliant garb of red, purple, green, and blue. But you know you are hearing the jewels of the marsh.

By late afternoon I have seen black ducks and mergansers. I've spooked a great blue heron from the shallows. It flew up the stream on slow, measured wingbeats. I've seen a muskrat swim by with its head partially out of the water, etching long Vs of wavelets onto the tranquil stream. I've caught sight of a deer near the shore, walking off into the mixed spruce-maple woods. There is recently chewed wood near beaver lodges along the shore, and you expect to see a beaver or an otter at any moment. Maybe even a moose.

When the sun begins its downward descent toward the horizon you might, out of the silence, be startled by the high, clear, tentative peeps of a spring peeper. This tiniest of frogs makes a sound like that of bells—clear, ringing, and pure. Neighbors join in, and immediately ensues a cacophany of shrill notes from an oxbow to your right, to be joined by another chorus to your left, further up or down the stream.

Now there is also a deep, rasping call that is repeated monotonously every few seconds. It sounds like a monstrous wood-boring beetle larva in a pine log, taking successive long and leisurely bites. As with the peepers, the calls of one individual stimulate a neighbor to join, and a local chorus picks up in volume as a distant one drops out. It is the mink frogs, one of the earliest breeders along with the wood frogs and the peepers. Spring has truly arrived now, and the rush of the migrant birds is on. Untold millions of them crowd the skies every night, traveling to these breeding grounds to take advantage of the newly emerging life.

May 4
SMELTING

There are hermit thrushes all through the woods already, and white-throated sparrows as well. Myrtle, black-and-white, and black-throated green warblers are back, but the others are not yet here. The kinglets and winter wrens sing exuberantly, and the tree swallows are still fighting over the nest boxes. This morning I saw a junco with its bill full of dry grass fly to a sedge tussock at the edge of the raspberries, disappear under it, and then come out empty-billed.

The smelt now run up the streams to spawn, and all true Maine outdoorsmen and women pursue these little silver fish at night. Smelting is a social event. The pain of wading in icy streams at night with a long-handled fishnet generally is made tolerable by the consumption of generous amounts of liquid painkiller. When the patient starts to hold up the little finned creatures and swallows them live, it is usually a sign that the medicine has been effective and sufficient.

Bill and I were going smelting. We'd settled it on Sunday evening over the phone. But when I got to his house he was nowhere near, and Millie, his wife, didn't even know he had made any plans.

"Did you talk to him Sunday night about it?" she asked.

"Well, yes, I did."

"We were having a party then, and he'd been having a few." No doubt.

"He must have forgotten," I ventured. But I doubted it. Billy wouldn't have forgotten about going fishing.

A neighbor boy of about 12 was shooting baskets into a net at the Adams' parking lot that doubled as a basketball court. With a friendly smile and flashing dark eyes, he tossed me the ball and we dribbled and shot for three-quarters of an hour. My movements were beginning to get smoother and I was feeling satisfaction from a few "swishes" as opposed to mostly air balls.

Bill still was not to be found. "I don't know where he is," Millie repeated. But she was anxious for him to return, too, because she had some work for him. "I'll make some calls to track him down," she said. Finally she came out of the house. "He was at Leona's [his mother], borrowing some bush-cutters. He'll be right here." Leona lives a mile down the road.

Bill's memory was perfectly intact when he arrived. "Just a second— let me grab my net and boots, and a six-pack." I shot baskets for another 15 minutes. Finally he came out of the house and we were off.

"Boy, was I glad to see you," he was chortling. "Millie had plans for me. That's why it took me a while to get out of the house. I told her it wouldn't be polite to leave you after you'd driven all this distance out here. She agreed. So I got points for being polite," he chortled loudly as he cracked the first can, took a huge swallow, and let out a burp of satisfaction.

We drove along in silence for a ways, well contented, and happy.

"Do you know how to gauge a quart?" he finally asked, and without waiting for an answer continued, "Two quarts is the limit. I've seen it where you can get your limit every time with one sweep of the net. You'll pull the net up, and the smelt will be spilling out all over the top. You'd have to get someone to jump in the water to hold up the rim of the net to keep them from spilling." No, I didn't know how to gauge a quart. But I'd never had the need to. Maybe I didn't know the good places.

There is a smelting gossip line. Bill had plugged into it, because his cousin Jake goes every single night about this time of year, and Jake in turn knows others who do, too. Together, they monitor the smelt runs in all of Franklin county. I figured it's like the information network that ravens have about where there is a good moose carcass. By yourself, you'd have no idea where to go to find where a smelt run might be on any particular week in this vast wilderness. You have to follow others.

"Where are we going, anyways?"

"I've got the directions here." He pulled out a crumpled piece of paper. "I talked to Jake over the phone while you were shooting baskets. Let's see . . . we go to Kingfield, then go past—and then two miles up this dirt road, and then park where all the other cars and pickups will be. Then just follow the beaten path. Did you bring a flashlight?" I hadn't. But the moon was nearly full, and the sky was only slightly hazy.

It was getting dark as we drove through the long, unbroken woods toward Kingfield. A moose started to cross the road in front of us. Bill doesn't have a bumper sticker saying "I brake for moose," but he braked anyway. Most Mainers just take it for granted that any critter over half a ton has the right of way.

The animal was molting its winter coat, showing huge bare patches of skin. It stood in the middle of the road and didn't move. At last we drove slowly around it, and then it turned back into the darkness of the second-growth forest of firs, red maples, and white birches.

We stopped at a small package store in Kingfield to get directions for the dirt road. It was an old logging road, and the ground was torn up from many vehicles that had spun in the mud. "A good sign," Bill said. "Smelt are only found where there are tire tracks."

As we continued, we finally saw where many vehicles had pulled off to the side, where they had spun to turn around in the bushes, and where hordes of people had tramped a day or so earlier. This was the place all right.

But there were no cars. "No smelt," Bill said.

"But you *said* you could never tell exactly when they'll run. Maybe they're running tonight and nobody knows."

"If you say so." Bill doesn't like to argue.

So we picked up our nets, smelt buckets, and beer, and followed the tracks. The trail could not be missed, even at night. There was tall timber all around, and the moon shone pale from above. We came to a few patches of snow. "The smelt run when the last of the snow melts," Bill said. This was more like it. I felt encouraged.

The path led down a steep hill, and soon we heard the roaring of a brook. But there was no yelling as there normally would be at a good smelt run. Nevertheless, we still beat our way through the underbrush along the brook in the dark, dipping our nets here and there. My net never felt heavy, but just in case I occasionally held it up against the moon to be able to see what was in it. We never saw a single fish.

Bill reassured me that the word *had* come down that they had been running here "Saturday and Sunday." It was Tuesday now. "They were either just starting or just finishing then, and I think it was the latter. The people here don't let on when the *real* run is. They'll put out the word that the run is on, just when it's near the end. That way they get more for themselves, plus they get others to tell them about *their* runs . . . To really do smelting, you have to be out every night till about 1 AM." I wasn't ready to "really" do smelting.

On the way back we stopped again at the place in Kingfield to pick up another six-pack. The little old man with the visor cap was still sitting by the cash register in front of a six-inch TV screen, as he had been when he had given us directions. He now asked how the smelts were running. We told him. He allowed a thin smile and cackled, "They's been sum a few days ago."

May 6
WOODS METAMORPHOSIS

There was rain last night, and with it has arrived a magical meta-morphosis. Overnight the hills have turned a startling new color: light pea green. Almost all of the trees except the ash are now unfurling their buds. The birches also are now shedding yellow pollen from their catkins, and entire serviceberry trees overnight have all flushed out in contrasting white.

The woods are alive with bird song. Today I hear five new warblers that until now had not returned: ovenbird, magnolia, parula, Black-burnian, and a water thrush down in the lowlands by the brook. The others—the myrtle, black-throated green, black-throated blue, black-and-white, and Nashville—continue to sing vigorously. I expect still nine more warbler species to appear here and in surrounding areas. By knowing their voices I am, in a sense, creating these fantastic jewels of the avian world, because if I did not recognize and name them, they would for all practical purposes not exist. Now when I hear one series of "tseets," my mind can visualize an intricately and beautifully colored bird in its nuptial plumage, as if he were perching directly in front of me; with another, slightly different series of sounds, I see an entirely different bird.

Along with the image of each bird, my mind also associates its nest. The Nashville warbler's is artfully tucked into a mossy hummock on the ground. The Blackburnian's is built of rootlets and twigs on the end of a limb in a tall red spruce. The redstart's is camouflaged with lichens in the low fork of an alder. The magnolia's is built in a young, vigor-ously growing balsam fir, whereas the ovenbird's is on open ground, tucked into a cover of fallen leaves. The black-throated blue's is a foot or so above the ground in a broad-leaved bush. Each of the 20 warbler species occurring near here has a unique nest, each in its own specific site.

Undoubtedly this specialization is due to the evolutionary selective pressure of merciless nest predation. The nests that survive are those that are best hidden—those in the least likely locations. A bird that nests where no others do has a good chance that no predator (squirrel, blue jay, crow) will be searching there, because all predators search on the basis of previous success.

The junco is now no longer carrying grass into its nest, but instead big billfuls of deer hair for the lining. A female robin is pulling up billfuls of dry grass in the back of the cabin. The swallows have sorted themselves out among the nest boxes: they have taken over three of them, a queen bumblebee has moved into a fourth, and a pair of chickadees is settling into the fifth. A male goldfinch in yellow and black nuptial plumage sings vigorously from the white birch by the cabin. I'm surprised, because these birds do not nest until three months from now, synchronizing their nestling production with the appearance of the thistleseed crop.

May 7

Inexplicably at dusk, the phoebe rose up into the air and sang excitedly in a different song, one which includes many rapidly strung-together "fee-bees," plus other sounds. Then he came back down. In previous years, the phoebes would have had small young by now; they are one of the earliest nesters. But this bird's mate had not returned this spring.

The woodcock only rose once to display in the sky, then spent the evening and dawn on the clearing—not high in the air. Is it finally running low on energy, or is the mating time over? The females are now each sitting on four beautifully colored eggs in some nest artfully concealed in brown leaf mold. They will sit there patiently, blending almost perfectly into the background.

There was a very slight breeze as I roasted potatoes for supper in the outside stone fireplace. A tiny young caterpillar came flying by, presumably suspended by a strand of thin, invisible silk, its skyhook and wing. Caterpillars use silk as escape ladders that they drop from trees in order to escape predators, and then they reel themselves back up to their food supply by rolling the silk line into a ball, shortening it as they climb back up. This is akin to how young spiders fly in the fall. Some spiders, of course, use silk to make their aerial nets to catch prey. Others use it to reinforce tunnels in dry soil, to make trapdoors, and to package their eggs. Many birds use spider silk for anchoring together materials in their nests.

Caddisflies use silk to make their cases, as well as exquisite fishing seines for prey drifting in the water. Some even catch microscopic particles such as bacteria in fine-mesh seines. Many geometrid moth caterpillars use a strand of silk to hold their stiff, stick-like form at an angle from a twig all day. This material aids their camouflage by saving them the muscular effort of keeping their body still all day, so as not to reveal their presence to caterpillar-hunting birds. (They move and feed at night only.) Similarly, many butterfly larvae use a single strand of silk across the midline of their body to hold the emerging pupa at an angle to their attachment site, while the tail end is secured to that site with sticky silk as well. Moths of innumerable kinds use silk to make solid cocoons, some of which are hard enough to protect them from being eaten by birds. We, in turn, use the silk from these protective boxes to make clothes, and even parachutes for warfare and for fun.

The little caterpillar that swung by in the swirling warm air near my evening campfire was anchored in silk, in order to fly. But its silk parachute, like many good things, was invisible to me.

May 14
FISHING

The chestnut-sided warblers finally returned to the clearing, coming on May 11, which seemed late. On his first day here, the male alternately sang and foraged for insects among the blooms of a pin cherry tree, where many queen bumblebees were searching for nectar and pollen.

From the woods beyond, I also now hear for the first time the

Opening buds

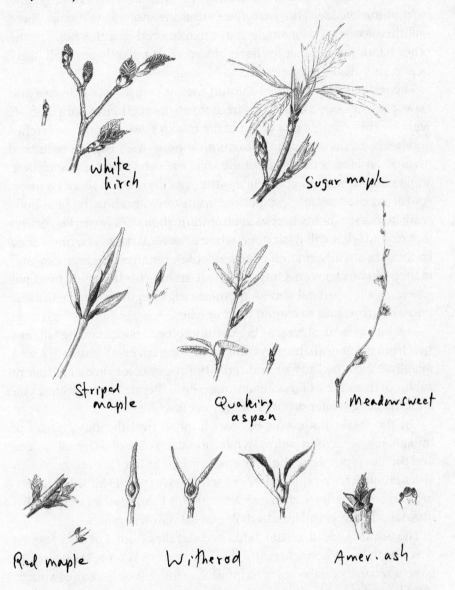

white
birch

Sugar maple

Striped
maple

Quaking
aspen

Meadowsweet

Red maple

Witherod

Amer. ash

Canada warbler, least and crested flycatchers, and rose-breasted grosbeak. Curiously, I have so far seen no Maryland yellowthroat here, although they are numerous in the bog.

My biggest surprise is the sugar maples. At this time of year, the trees usually are light yellow from millions of small flowers that hang and twirl in the breeze. This year there are none, not a single one. There will therefore not be a single sugar maple seed cast this fall. On the other hand, the red maples have bloomed abundantly and will likely bear a very heavy seed crop.

The red oaks are starting to unfurl their greenish yellow leaves and flower tassels, but the flowers are not yet open. Their color contrasts with the shiny, bright pea green of the quaking aspens and the birches and beeches, the silver of the bigtooth aspens, the rusty red of the red maples, and the brown gray of the still bare ash trees. Here and there some trees seem dipped in white—the pin cherries and serviceberries. And along the roadsides, some wild apple trees are showing pink buds ready to burst. Chokecherries are unfurling their dark green leaves and flower buds that will flash open white in several days. The deciduous larches are already clothed in new needles, but the evergreens are still in their old black green. Only the firs are starting to show their first light green, as the terminal shoots on their twigs begin to pop out of the buds, to grow, and to extend their needles.

The hillsides are at least as beautifully colored now as in the fall, in a spectrum ranging all the way from white through every shade of green imaginable to the browns and reds. But these colors are much more subtle than in the fall, and even more ephemeral. They astound you after the long winter drought of grays and white.

On the forest floor, you have to hunt to find the tiny specks of brilliance—the lemon yellow, white, and deep blue of different violets, and the deep purple of the trilliums. It is like finding jewels, however, and a contrast to the splashes of color that cover whole hillsides, bleeding into one another as if an overzealous child had been let loose with a thousand brushes and hundreds of pots of different paints.

The sounds of all of the birds are dazzling, but I'm at a loss to describe them. When you read "blue" or "red" or "green," you already have a picture of each color in mind that these words can reawaken. But I have no words with which to conjure up in your mind the lilting, lisping song of a black-throated blue warbler, nor with which to give you even a taste of the vibrant, energetic refrain of a winter wren. These sounds come from another world that must be experienced to be felt. There is a limitation of vicarious experience, which reminds me of why I came to these woods in the first place.

The phoebe yesterday inspected the log under the cabin window where its nest was last year. It then went into paroxysms of calling, while flying back and forth from that spot on the log and the trees near the cabin. But it is still alone.

Bill called and left a brief message on my answering machine: "Come at 6 PM. Bring fishing pole and bucket. White perch at Maranicook Lake." It was the signal I'd been waiting for.

I had heard stories for years of how you could stand on shore and pull out one perch after the other, "just as fast as you can bait your hook." Having myself fished for white perch all these years only by lazily dangling a line over the edge of a rowboat in the middle of Pease Pond, and hoping for luck (and not always having it), I was eager to see the real thing. "There will be kids and their grandpas, guys, girls; *everybody* goes, and the shores are *lined* with people, all pulling them in hand over fist. It's quite a spectacle."

I got to Bill's house in Jay on time, and his two sons, Cutter and Chris, were ready, too. They loaded their buckets and poles into the trunk of their car and I added mine. We were off in no time, this time, stopping at the general store in town, right next to the I.P. paper mill, to buy a carton of night crawlers and a six-pack. Then we traversed hilly Main Street past the frame houses and not-too-elaborate storefronts of both Jay and Livermore Falls. In less than five minutes we were on the woods-bordered road to Maranicook, and I was hearing more fishermen's tales of the fabulous perch runs there.

Turning right at the Readfield general store with its sign of peeling paint, we came down a hill, and the lake spread to our left as the road crossed a small cement bridge at the pond inlet. This was the place. It didn't look good, though—there were only three cars parked alongside the road. But it was still light out. "They'll run just before dark," Bill said.

Several young boys and girls were scampering about just playing. But two young men and one man who looked older were *fishing*. We checked what they'd been getting. So far they had only caught sunfish and yellow perch, which are considered "trash" fish here in Maine. "They eat them in New Hampshire," Bill said, but I guess that doesn't count for much.

A sunfish has green sides with shimmering flecks of gold that reflect light like little mirrors. Its orange and blue face is marked with a bright red decoration at the edge of its gill covers. Yellow perch shimmer like gold, and have greenish vertical bands. White perch glitter like silver. Large black pupils stare from their huge silver eyes.

The old man said, "I don't think they'll run tonight. I've been here every night this week and haven't caught one yet."

"I tried it last week and didn't get anything, either," Bill admitted.

White perch w.
8, 1, 1 + 3 erectile
spines. Don't look.
silvery – esp. eye

The old man reeled in his line, threaded another worm onto his hook and cast again up the stream, letting the line with a red bobber drift down toward the cement bridge. Suddenly there was a small "plop" as the red bobber disappeared under the water in one smooth sucking motion. Only circles of tiny concentric ripples remained on the glossy black water.

We watched and felt the excitement as the monofilament line sliced through the water. The old man's pole bent as he reeled, and when he was about to pull the splashing fish out onto the bank, we noticed that it was a bass. It is illegal to keep bass at this time of year, but he was saved the trouble of releasing it because it fell off the hook on its own.

The near-catch energized us, and we now cast out our own lines.

Nothing bit. No bobbers were yanked under the cool dark water, nor were any even jiggled.

The sun approached the horizon, and golden reflections played on the black and blue wavelets of the stream. The dark silhouettes of a silent pair of loons out on Maranicook Lake, on the other side of the bridge, caught my eye. Then a yellow warbler came by, searching for its last evening snack of mayflies in the willow thicket off to our side. We heard a few resounding "thunks" of a bittern calling from the bog above the inlet, and also the liquid twitter of a waterthrush. The pine trees toward the setting sun became stark black sentinels. Still no bites. I put down my fishing pole and walked over toward the old man. Meanwhile, the two boys had caught six yellow perch, one sunfish, and finally even one white perch.

"I don't think they'll run tonight," he said again. Bill joined us, too. "They haven't been running here the last three years," the old man finally admitted, "and I've been fishing here for 70 years . . ." After another little pause he added, "And I wasn't a toddler then. I was 22."

There was no moon out, and it was pitch-black when I left Bill and the boys and walked back up the path to the cabin. I had no flashlight. Suddenly in the woods next to the path I saw a pea-sized ember of a white, almost bluish neon light. Unlike the firefly's light, it shone steadily without blinking. Gently picking up the small ember, I held it between my thumb and forefinger. It felt soft, but I could not determine its form as I cradled it in my hand and watched its light, then walked on.

After a hundred yards it disappeared from my hand. Had the light gone out? That did not seem likely, because the light had been steady and unwavering. It must have fallen out of my hand, despite my trying to clutch it tightly.

I walked back, and there it was again, glowing in the middle of the path, and now moving along slowly like a snail. I picked it up again, and now realizing that it was an insect, I held it tightly in my fingers until I could confine it in a film canister at the cabin. The next morning I examined and drew it. It was the *larva* of a firefly (which actually is a beetle). In the adult, the light flashes function as mating signals. But in the larva?

I also drew some of the fish. One by one I chopped up some yellow perch and fed it to the ravens. They were *wild* for yellow perch. I had never seen them so excited on their feed. So it occurred to me to try a taste test for myself. I cleaned the white perch, a yellow perch, and a sunfish. All were close to the same size. I rolled them in flour and then fried them in oil. At first I ate only a small piece of each fish. They all tasted the same. Was I being biased to try to make the "trash" fish good? I then mixed the pieces all up, eating chunks of fish at random. I still could not detect any difference.

Among the crowd of local fishermen, I was the only one who was eating yellow perch and sunfish. All of the kids there fishing at the bridge had learned from their parents and associates how to tell a "good" fish from a "bad" one. And they would teach the difference to their kids, in turn. They all knew, without even tasting, that the yellow perch were not good. They had waited for the *white* perch, and they had all gone home empty-handed. I have now learned *not* to know any better.

May 15

WEEDS IN HUCKLEBERRY BOG

Like yesterday, there is frost on the ground at dawn, but by 7 AM the rising sun has melted it to dew. The sun also touches the apple blossoms of the tree by the cabin, and they are now beginning to open in a blaze of pink. A male ruby-throated hummingbird appears on cue, hovering among them. It is a perfect day to get reacquainted with Huckleberry Bog. I want to hear again the yellowthroat, the palm warbler, and the olive-sided flycatcher.

To get to Huckleberry Bog you drive along a tarred and potholed winding road, then walk in along a logging road. To my great surprise I notice now that the beeches *as well* as the sugar maples also are not blooming at all this year, and I eventually examined thousands of trees of both species. Everywhere I still see beech seed capsules on the trees from last year's crop, so last spring the woods must have been ablaze with their flowers. This year there is not a single beech or sugar maple flower in sight, although the leaves of these trees are almost all out and the new twig shoots are growing vigorously. It seems a strange spectacle to find two dominant forest trees not blooming at all this year, whereas in other years the hillsides were covered by their bloom. The entire woods, however, are still tinted red from the red maple seeds that are almost ready to be shed.

The "decision" of two tree species not to bloom was not arrived at this spring. It already had been made by last fall when the buds were formed. It may even have been made much earlier.

Yes, trees do decide; they register information from the environment and they respond to it in appropriate ways that are related to survival. If they do not produce seed now, it must somehow result in producing more surviving offspring in the long run. How?

Last year was, as they say, a good "mast" year—a bumper year for both beechnuts and sugar maple seeds. Many of these seeds are now sprouting; I've counted up to ten seedlings per square foot in some places. Bears; deer; mice; red, gray, and flying squirrels; grosbeaks; blue jays; and seed-eating beetle larvae all were growing fat and consequently reproducing. However, last year all these seed destroyers were still at low densities and there were seeds enough for all to eat, with many left over for producing seedlings.

If the trees produce such bumper crops several years in succession, then all of the seed predators will dramatically increase in abundance and will then converge on this reliable food resource. Organisms reliably fed also reliably reproduce up to the limits of the *bottlenecks* of their food supply. By cutting off their predators' food supply *this* year, the trees have narrowed the bottleneck—by not reproducing now, they have helped to ensure the survival of embryos to be produced in *future* years. In effect, they are preventing the population explosion of their predators before it occurs.

The distribution of blooming (and hence seed production) to occur at widely spaced yet synchronous episodes is enforced by penalties. Any beech tree blooming when its neighbor doesn't will waste energy, because its flowers will remain unpollinated and thus useless. Even if this maverick tree should go so far as to self-pollinate and produce seeds, then most of these would be wasted, too. That is because the now hungry seed-eaters from all around would converge on the one bountiful seed-tree. Their appetite for this tree's seeds could not be diluted by those of its neighbors, and they would leave no seed uneaten. Again, all reproductive effort would be wasted. But if the *whole* forest floor were suddenly covered with seeds, as happened last fall, then although a few seed predators would enjoy a temporary feast, their feeding on any one tree's seeds would be nevertheless dispersed among many trees. Therefore, a tree that manages to produce seeds when most other trees of its species are producing theirs will ensure the greatest survival rate for its seeds.

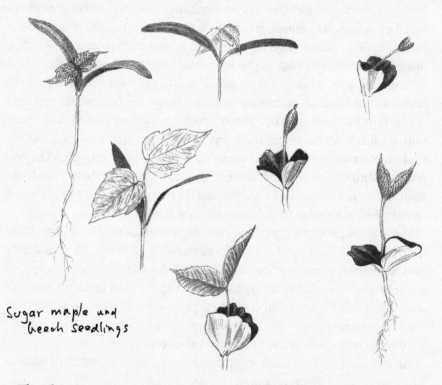

Sugar maple and beech seedlings

The dominant trees of the Maine uplands are the beech and the sugar maple, and it seems intriguing that *both* of these major hardwoods with large seeds would relinquish flowering in the same year, whereas most of the trees that either occupy different habitat or do not have large enough seeds to be animal fodder, are still blooming as they always do. (The small-seeded trees *never* miss a year.) It appears that the two large-seeded species are cooperating to achieve a common objective, a principle called coevolution. The existence of coevolution has been debated by academics ad nauseam. The argument seems odd, because everyone agrees that the selective pressures provided by other organisms on one another likely are *at least* as strong as are the selective pressures provided by the physical environment. A beech tree that produces seeds when its neighbors do does not care whether those neighbors are other beeches, maples, or some other species, so long as the seed predators are the same for all.

The bog at this time of year is alive with the "witchety-witchety" refrain of Maryland yellowthroat warblers and the trilling of palm warblers. Most of all it resounds with the loud and penetrating call of the olive-sided flycatcher, sometimes described by woodsmen as a melodious whistle of "whip-three-beers," which probably reflects more what's on their minds than what the bird sounds like. I am relieved to hear all three species, and many more. When I hear all the right birds, then I know all the world is well.

The edge of the bog starts abruptly at the foot of the hill. Here you enter thickets of gray and brown speckled alder, winterberry, highbush blueberry, and withered and privet andromeda. Last year's growth of sedges lies tan and matted on the mud, with sharp new six-inch spikes of bright green shooting up through it in dense patches. A few yards further are the low mats of leatherleaf, still holding last year's now leathery brown leaves like little mouse's ears. Their twigs are replete with rows of tiny white bell-like flowers, and you hear the gentle hum of bumblebees. Dense, low bushes of rhodora, still leafless but with swollen purple flower buds, will be ready to burst open into flaming blossoms soon, when the leatherleaf stops flowering.

Interspersed among the rhodora are a number of other ericaceous plants. I particularly notice the low shrubs of Labrador tea, whose rough, dark green leaves with light brown fuzz underneath seem unchanged by the winter, but whose globular buds at the tips of the twigs are pregnant with snow-white umbels. The buds of the sheep laurel look unchanged since winter, and their small pink flowers will open later, near the end of the flowering queue that sees each species pollinated in turn by the same cohort of bumblebees. Each species' bloom ticks off the time of the season like a clock, all to their mutual benefit, and to the mutual exploitation of the bumblebees, who in turn rely on them for food—another example of cooperation achieved through self-interest that is a dominant theme in the evolution of life on earth.

All of these low perennial shrubs are embedded here in reddish green sphagnum moss. A few yards further into the bog, the moss forms a solid, waterlogged carpet. You sink in and feel the coolness of the brown bog water. The moss here is crisscrossed with thin cranberry vines bearing tiny oval leaves and the tart purple fruit from last year that was edible in the winter and still is edible now. The bog laurel reaches inches above the moss, showing old yet green leaves and twiglets topped with bright pinkish purple flower buds. They sit fully formed and ready, waiting along with other buds for precisely the right time to open and offer their pollen and nectar to the bees.

Here, too, are patches of bog rosemary, whose narrow, pale bluish green leaves have pure white undersides. The buds hang small and pink, to become little bell-like flowers soon after the bog laurel is done, but before the sheep laurel has bloomed. There also are sundews and pitcher plants below, and stunted black spruce and larch trees above.

This delicate mosaic of many greens, mixed with dabs of bright color, is a community of plants and animals that has been in existence here virtually undisturbed since the glaciers retreated some 10,000 years ago. It is a sacred place to me—a place for refreshment and wonder.

I might have left refreshed on what otherwise would have been a perfect day. But taking a slightly different route out of the bog than how I had come in, I was suddenly shocked to see white styrofoam chips spread in a compact pile at my feet. The chips were of the indestructible kind that is used to cushion coffee mugs and other delicate, precious mail-order items, when crumpled newspapers just won't do. Examining the pile more closely, I noticed that it was surrounded by a now broken black plastic bag. In the middle of the pile was a generous portion of non-bog soil, out of which I could see several dried cutoff stalks of a plant I judged to be non-native—a weed, literally. A further examination of the environs turned up 35 other similar mounds of trash.

Clearly, the weed cultivator had chosen this spot because he or she judged it to offer the least attraction to anyone. It seemed to me, however, that it was one of the most beautiful spots imaginable—one that should be inspected regularly and in great detail. I shall erect a sign here. It will say, "This is a pristine bog. Don't make it look like a dump."

May 24
GRADUATION

I awoke while it was still dark, at 4 AM, to the loud, stirring song of a whippoorwill. It sang a few refrains behind the outhouse, then once more further down the hill, and then I heard it no more. I fell back asleep, thrilled because I had not heard this song since I was a boy at the Adams farm.

The call united me with the past. The bird would always sound as it sounded now and as it has sounded since humans first heard it in the mists of time millions of years ago. And so it will sound when I hear it again.

I got up, refreshed, as it was just getting light, about 5 AM. The phoebe had been calling loudly for quite some time by then, and a red-eyed vireo had chimed in. Half an hour later there was still a little frost on the ground, and the sprouts of asparagus I had planted last year looked wilted and dead. The apple tree in front of the cabin was in full bloom, and small white wild strawberry blossoms were sprinkled throughout the short grass under the lone white birch. A chickadee was pulling chinking out from between the logs of the cabin in order to line its nest in the bird box, and I did not begrudge it a single thread.

*   *   *

Yesterday I came back from the city of Burlington, having gone to attend my daughter's college graduation and the hooding ceremony of my doctoral student, Brent Ybarrondo, and to visit with Stuart. Before my sojourn in the woods, I had never before attended any of my own graduation ceremonies, nor any of my students'. The ceremonies always were held several days after the last exam, and within hours after that last test I was always *gone*—off to the spring woods. To me, ceremonies had always seemed a waste.

Now I wanted to be there. I had even volunteered to help. My job was to assist in the lineup of the biology-zoology majors, which included not only student friends of mine, but also my daughter and Dave, her boyfriend. Erica had earned the prestigious Lyman Award in zoology and a Phi Beta Kappa, as well. I had never cared much about those things, either, but now I felt immensely proud.

I was enthralled by the many smiling faces around me. There were hugs, reminiscences, small talk about the past ("I *loved* Winter Ecology"; "You have a great place there"), and promises to keep in touch ("I'm going to drive across the country"; "I'm going to Patagonia"; "Remember to write . . ."). I only had to confiscate one bottle from a student, who objected because he hadn't finished it yet.

The marshal came to lead the students from the assembly points at the tennis courts, where we had lined them up, out into the gym where all the friends, parents, and significant others were assembled. I had stayed too long chatting with students, and had missed the faculty procession that had gone in first. No matter; I belatedly squeezed through the throng. The other professors were not hard to find near the front of the gym. The dean gave a speech, and so did many others. The retiring professors were each honored with a short speech praising their deeds, and then the students marched up in lines by departments. Alphabetically, one by one, they had their names called and received their diplomas. Parents and friends crowded near the podium to take pictures. Faculty with children graduating were even allowed onto the stage to congratulate their son or daughter immediately after he or she received the final, official sign-off. I, too, went up, and hugged Erica, feeling glad, and feeling the gladness of everyone else.

Afterwards, in the huge throng milling about in the spring sun outside the gym among fresh lilacs and apple blossoms, I again met Kitty, Erica's mother and my former wife, and her parents, Cos and Mary. They had come all the way from California to witness the occasion.

Later on, I headed over to the Ira Allen Chapel, where the doctoral degrees were being awarded. My student, Brent, had finished a superb thesis on the diving physiology and behavior of a water beetle that has solved the problem of breathing under water. Brent had devised a new and ingenious technique for measuring how much air a beetle takes down on a dive, and through his long and detailed observations and measurements, he had made several discoveries. Brent's work was not a professor-assigned project—it was uniquely his, as any doctoral work must be. A Ph.D. degree requires an original discovery, and since such things cannot be guaranteed, to embark on a doctoral program is a big risk. I was delighted to celebrate Brent's accomplishment, already recognized a year in advance, when he had received a job offer to teach at a college in Colorado, precisely where he wanted to live. Now he had come back for the formality of receiving his velvet doctoral hood and his diploma. His wife, his mother and grandmother, and two of his close friends had traveled from the other side of the continent as well. Here, finally, was reason for me to attend a graduation ceremony. I proudly put the yellow and black hood over his black robe when his name was called, and we walked together onto the stage in front of the assembled crowd.

But the next day, I returned to Maine . . . with two younger students, a new raven, and my son.

I'm still catching mice. I have a new young raven, and he likes them raw. Since I don't have a refrigerator, I cannot store up enough of them for a good mouse fry. However, I have saved several of these mice for entertainment.

Having a nine-year-old boy with unbounded energy and curiosity, every day, all day, is a challenge perhaps akin to being a social director at a summer camp. In addition to guiding, cajoling, threatening, and deflecting the inexhaustible youngster's energies into socially and materially nondestructive channels, I've also got to deal with the aftermath myself.

"Stuart," I said, "let's lay one of these dead mice out here on the ground. Maybe we'll see a burying beetle come bury it!"

We put a mouse down onto the hard-packed sawdust in back of the cabin, and I explained that the sexton beetle, if it came, would be clad in elegant deep black with bright orange stripes, a regalia like a graduation hood, I thought. It would smell the mouse from perhaps half a mile away, flying upwind very much in the manner of a bumblebee. Finding the mouse, the beetle would bury it in the ground; if it was a female, she then would use the decaying meat to feed her young after laying eggs on it.

We checked an hour later. The mouse was gone! I felt magic, although I don't believe in magic. A beetle must be near, I told Stuart, and he should find it.

A few minutes later he ran in yelling, "Dad, Dad—I *found* one!" Stuart's enthusiasm had been kindled. He had discovered where the mouse was buried, and a sexton beetle was with the mouse. He instantly transferred the beetle to his killing jar (filled with ether) to add it to his growing bug collection.

To see a sexton beetle at work before it had finished its task, I now put the excavated mouse onto harder ground. Stuart, having been made a believer, sat down next to the mouse to wait for the next one, as I had hoped. I went inside the cabin. I had seen this part of the show before.

For twenty minutes, my son was as quiet as I've ever seen him. But suddenly he burst in the door. "Come *quick* Dad, one came—it sounded just like a bumblebee and I saw it with its wings still open just after it landed. It has gold fuzz on its back just like a bumblebee!"

We both sat on a log of the woodpile, watching the beetle alternately excavate soil from underneath the mouse and at times stop motionless with its abdomen stuck up into the air.

"It is calling a mate with scent," I told him.

Sure enough, in only a few more minutes still another beetle flew in, and then the two of them worked together under the mouse. It was bonding on first contact, no doubt due to the difficult task they now shared.

The two dug and heaved. Loose dirt was thrown up all around the edge of the carcass, as the two dug out of sight beneath it. But the heaving of the dead mouse betrayed their presence. Occasionally one of the handsome creatures crawled over the white belly of the deer mouse, to then again resume work on the other side.

"What are those tiny crawly things on the beetle?"

"*Those* are mites that eat fly eggs."

"Do the mites harm the beetle?"

"Not at all. See those big, shiny green blowflies? They are laying eggs on the mouse. Within two or three days, maggots from those eggs would eat up the entire mouse. Now these mites are going to eat the fly eggs, and then the *beetle's* young will eat the mouse."

"I see!"

Stuart sat down and watched the beetles for well over two hours. I'd never seen him so absorbed. Even after I stopped watching the burial with him, he continued to give me progress reports. I didn't have a TV to kick in, but if I had had one, now would have been the perfect time. Nature provides the greatest entertainment on earth, and we had watched a flawless *live* performance that had been rehearsed on warm summer days and evenings for at least 60 million years.

I have always associated summer days with sweaty work lifting hay bales onto a truck, with hoeing and weeding long rows of corn and beans, but also with the ecstasy of swinging on a rope from high up the bank into a shaded stream.

My anticipation of the cool water is now enhanced not by work in the fields, as when I was a farmer's apprentice, but by work in the woods. I'm already replenishing the woodpile for next winter.

One of the many local swimming holes that also has a swinging rope is on the Sandy River, by the bridge near the Farmington Diner, just as you come into town. The river here is broad and smooth, and the clear water off the hills flows over a wide sandy bottom. The water shimmers yellow from the sand below, and reflects green from the trees along the shore. Silver maples grow atop the 15-foot-high banks. Their trunks have scars where the ice ground off sections of bark during the spring thaw. They lean far out over the water, and far out on one of them hangs a rope. Here is a gathering place for kids of all ages.

The distance down to the water is sufficiently high, and the rope sufficiently long, so that you get a good ride when you launch yourself off the bank. You make a downward swoop until you are barely skimming the water when you attain top speed. Then the momentum and the rebound from the tree take you far out over the river, and you go up, and up. You have to go where flight takes you, and you have to know when to let go. Usually we let go at the instant when we're stationary again, just before the pendulum begins to swing back toward shore, when we're out at the farthest and highest point above the river. *That's* the time to let go. For a second you float through the air and then . . . I've let go on Adams Hill already, and the cool immersion invigorates, as I plunge all the way to the solid, sandy bottom.

Like the indigo bunting and the phoebe that live here, I've traveled widely. I've lived in California, Africa, New Guinea, South America, and the High Arctic. But I've come back to the hills of western Maine. These are my favorite haunts, because this *is* home, where the subtle matters, and the spectacular distracts.

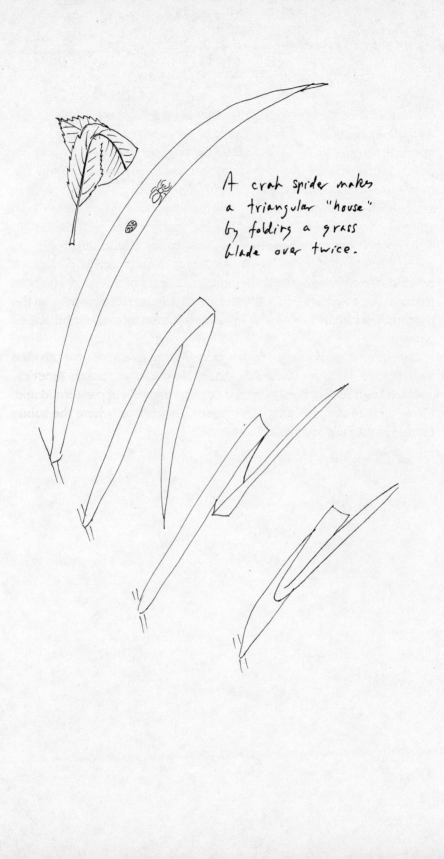

A crab spider makes
a triangular "house"
by folding a grass
blade over twice.